THE ORGANIC BABY BOOK

Tanyia Maxted-Frost has lived and worked in Australia, NZ and the UK. She won the Young NZ Journalist of the Year Award in 1984 and moved into environmental journalism in Australia in 1989.

Arriving in the UK in 1996, Tanyia became interested in organics when writing 'Focus on Organic Farming' for the National Farmers Union. In 1997 she co-founded the London Organic Food Forum, created and edited the Soil Association news service *Organic Food News UK* and co-organised 'The Future Agenda for Organic Trade', the UK's first international organic trade conference in Oxford. Tanyia also created the talks programme for The Organic Food & Wine Festival for two years.

A freelance journalist based in London for nearly four years, she wrote columns and feature articles on organic issues for many national consumer and trade magazines, and wrote two organic books—this one, and *The Organic Baby & Toddler Cookbook* (co-authored with Daphne Lambert), also published by Green Books. She now lives in Perth, Western Australia, and is Editor of *Homes & Living* magazine.

The
Organic
Baby Book

How to plan & raise a healthy child

SECOND EDITION

Tanyia Maxted-Frost

GREEN BOOKS

This second edition published in 2003
by Green Books Ltd, Foxhole, Dartington,
Totnes, Devon TQ9 6EB

Copyright © Tanyia Maxted-Frost 1999, 2003

Research for Products Section
by Kristen Steele, Clive Litchfield and John Elford

Cover design by Rick Lawrence

Cover photos of organic baby Tane Maxted-Frost,
aged 8 months, by photographer Michael Bassett

Typeset in Sabon and Stone Sans at Green Books

Printed in Great Britain by
MPG Books, Bodmin, Cornwall

A CIP record for this book is available from the British Library.

ISBN 1 903998 21 2

Disclaimer

This book is a general guide which explains the potential benefits of organic foods, and describes currently available products and services. Women who are pregnant, breastfeeding, or trying to conceive are advised to seek professional advice before undertaking any health programme, taking supplements, etc.

CONTENTS

Part Two

ORGANIC PRODUCTS GUIDE

Acknowledgements

I would like to thank all the experts and parents quoted in the book for their valuable contributions and encouragement to us all; all the people behind the organisations, publications and companies listed and featured in the book for working tirelessly towards a better future for our children—particularly those in the Foresight Association for the Promotion of Pre-conceptual Care and the Soil Association; Daphne Lambert of the Penrhos School of Food and Health, who kindly supplied lots of useful organic food, health and nutrition tips from her mother & baby course; Antony Haynes and Tania Smith, and Helen Pitel. Thanks also to my (mostly) understanding and patient husband Andrew; my enthusiastic and motivating editor John Elford; to Kristen Steele for all her work in updating the Products Section for this new edition; and to my son Tane for putting me on the path to organic motherhood and this book in the first place.

*   *   *

I have taken some liberty with the legal term organic to create the terms 'organic' baby, 'organic' parent, 'organic' parenthood and 'organic' breastmilk, as none of these can actually be certified as organic. The legal requirement for using the term organic for marketing or selling products or services in this country is certification by an approved certifier or overseas equivalent.

For simplicity, babies are referred to by the masculine pronoun ('he')—apologies to all the baby girls out there. As some readers will find certain sections more relevant than others, some of the more important information and advice is given in more than one place in the book.

For my son Tane
and all the other organic cherubs

Introduction

BABIES ARE TOTALLY DEPENDENT on us from conception onwards, and as parents their health and wellbeing is our responsibility. The food and drink we give them, and the clothes, toiletries and household products we put on and around them, can be beneficial, benign, or harmful to their health in the short and long term—perhaps even fatal. When we have a baby, it may be the first time that we actually stop and seriously think about what we're eating and using in our homes. Naturally, we want to protect our children, and make sure they are as safe as possible.

The good news is that there are now enough organic and environmentally friendly foods and products available in the UK to bring up babies safely and healthily, enabling parents to avoid exposing them to the many harmful toxins and undesirable additives commonly found in conventionally produced foods and goods. A comprehensive directory reviewing them all can be found in the second part of this book—and they are becoming more affordable all the time.

It is now possible to have a baby conceived by 'organic' parents, raised on 'organic' breastmilk and an organic wholefood diet, clothed in organic nappies and baby-gros, bathed in organic bodycare, and bedded in organic blankets on an organic mattress: a truly healthy, vital infant with a strong, unimpaired immune system able to withstand life's knocks—a baby given the best head start in life.

So 'organic' parenthood has finally arrived. It's an exciting time: by going organic as much as you can—and can reasonably afford—you and your baby are helping to pave the way to a brighter, healthier, less polluted future for us all.

I WROTE THIS BOOK because I wanted to share this good news with other like-minded parents, prospective parents, and pregnant mums. Our children have the right to eat nutritious food, free of artificial chemicals and genetically modified (GM) ingredients, and not to have their health compromised or damaged by these and other man-made toxins in everyday foods, household and consumer goods. It's our duty as parents to protect our own and other children from these very real dangers. We should all have the right to affordable, safe, wholesome food. That's why I wrote this book—while pregnant, breastfeeding and co-raising my own 'organic baby'.

I also wanted to support others working in this field: the people growing, making and selling organic foods and products, who do so out of their own strong beliefs, often with little remuneration; and the pioneering charities and organisations like the Foresight Association for the Promotion of Pre-conceptual Care and the Soil Association, who are doing so much to create a healthier nation, against all odds and with little money.

I sought, and have featured at some length, the opinions of leading and popular health experts. This is because I wanted to show you that such experts now recommend it as vital for your baby's health, and no longer simply just desirable. I also sought and have featured the experiences of many parents who have 'gone organic', to show you the difference it has made to them and their babies. I haven't included much of the standard pregnancy and babycare advice and information, as I found during my own pregnancy that there are already many excellent books which offer this.

Many of the statistics and facts used in the book are now widely accepted. They have been drawn from a wide range of reputable and well researched sources, including *The Shopper's Guide to Organic Food* by Lynda Brown, publications of the Pesticide Action Network UK, Foresight, Sustain, UNICEF's UK Baby Friendly Initiative, the London Food Commission's *Food Magazine*, *The True Cost of Food* report by the Soil Association and Greenpeace, *Our Stolen Future* by Dr Theo Colborn, Patrick Holford's *Optimum Nutrition* book series, the WWF-UK report *Chemical Trespass: a toxic legacy*, and substantiated national media reports by leading food and health journalists. Many of these sources are among recommended publications listed in Part Two of the book.

Finally, I hope this book can be a catalyst for change and introduce many more parents to the 'wonderful world of organic', and support those already converting or converted as a useful reference. I believe that the vital choice we make to go organic for our babies' health will enrich our parenthood, as well as our children's health and lives. I know it has enriched mine.

Tanyia Maxted-Frost

A few words from the experts

Maryon Stewart (founder of The Women's Nutritional Advisory Service, author of *Healthy Parents, Healthy Baby* and *The Phyto Factor*, and mother of four):

"Mother Nature, in her great wisdom, creates the environment in which a newly conceived baby can grow, and provides us with breastmilk, the perfect food to nourish the baby during its first months. However, research now shows that she needs a helping hand, as the nutritional shape that both parents are in—at least four months prior to conception—will determine not only the health and wellbeing of the newborn baby, but also its chances of health in adult life.

We all know it takes a female egg and a male sperm to make a baby. But did you know that both eggs and sperm may be adversely affected by inadequate diet, and by social substances like tobacco and alcohol, drugs and chemicals, and environmental factors? This relatively recent knowledge places a great deal of responsibility on the shoulders of prospective parents, for our general health, dietary habits and fitness level all play a major part in influencing the health outcome of our unborn children. Our nutritional state and our environmental conditions can to a large degree determine the risk of heart disease, stroke, diabetes, high blood pressure and cholesterol levels, as well as favourably influencing their intelligence and physical abilities!

That is just the beginning. We also now know that the type of diet our babies are fed during the first year of life will also play a major part in determining their health prospects. For example, a baby deficient in iron may develop anaemia and have growth and learning disorders. And babies who are fed the wrong type of food too soon, during the weaning process, could develop allergies in later life or coeliac disease.

By making sure that our eggs and sperm are in tip-top shape before pregnancy, that we eat a wholesome diet with plenty of organic food during pregnancy, and feed our babies with nutritious chemical-free food, we will be providing this new individual with the best possible start in life. Becoming parents is an indescribable joy. Knowing that you have taken every possible step to ensure your baby's well-being can only enhance that joy."

Dr Sarah Brewer (health columnist and author, and mother of Saxon):

"The shops are full of plump, juicy fruit and vegetables that have been specially bred for their colour, uniform size and ability to keep their appearance longer. Often, this has been achieved at the expense of flavour and nutrients, and with the help of a range of agrochemicals such as pesticides, weedkillers, fungicides, fumigants, growth promoters, growth retardants and fertilizers. These chemicals are applied regularly, from the time the crop is still in its seed form, during germination and throughout its growing cycle. Each non-organic apple, for example, has been dosed around forty times with up to 100 chemicals before you eat it.

These chemicals do not just lie on the surface of the produce, but are found beneath the skin and sometimes throughout the flesh itself. The full effects of most of these chemicals on our long-term health, immunity and reproductive system are still not fully understood. That's why increasing numbers of people are deciding to 'go organic'.

As a family we have eaten as organically as possible for two years and wish we had taken the plunge far sooner. It was much less complicated and less expensive than we had imagined. We belong to two local box schemes and obtain organic meat from our local farmer. Food is more flavoursome than before and we have all noticed a slow improvement in our energy levels as well as our overall health. I would highly recommend going organic for healthy children to any parent."

Patrick Holford (author of *The Optimum Nutrition Bible*, *100% Health* and *The Optimum Nutrition Cookbook*; founder of the Institute of Optimum Nutrition, and father of two):

"Organic food is absolutely essential for a baby. There are two good reasons for this. The first is that there is increasing evidence that very small levels of pesticides and herbicides—found in non-organic produce as residues—can act as hormone disrupters, effectively disrupting the early hormonal programming of a child which can affect its development, including its sexual development, later on in life. There is

> "A fertile soil means healthy crops, healthy livestock and last but not least healthy human beings. Soil fertility is the basis of the public health system of the future."—Sir Albert Howard (1873-1947), regarded as the first leader of the UK organic farming movement.

real concern that the big increase in hormone-related problems, not least of which are breast cancer, which now affects one in eight women, and prostate cancer which is predicted to affect one in four men by 2015, is associated with very early exposure to hormone-disrupting chemicals by eating non-organic produce in childhood.

Other conditions associated with exposure to hormone-disrupting chemicals by eating non-organic produce include polycystic breast disease, fibroids and endometriosis. A non-organic diet is also more likely to contribute to chronic fatigue than an organic diet. And we know without a shadow of a doubt that a lot of children with Attention Deficit Hyperactivity Disorder are sensitive to chemicals—this has been shown very clearly with certain food colourings, and given that there are now three and a half thousand chemical additives allowed in our food chain, it would make sense to cut that load down. Parkinson's disease is linked to pesticide and herbicide exposure, and eating non-organic produce is a contributor to Alzheimer's. So eating organic produce is the only way of guaranteeing that you and your baby are not getting these very powerful anti-nutrients.

Secondly, several pieces of research claim that organic produce actually contains substantially higher levels of nutrients—somewhere between 50–100% higher levels of minerals, on average, than non-organic. Most people are not aware that organic produce generally has 25% more dry matter. In other words if you look at plants as consisting of water plus nutrients, in non organic produce there is more water and less actual nutritional matter. So two organic carrots equals three non-organic carrots. If an organic carrot is 25% more expensive, it is still actually the same value for money in terms of actual carrot. The rapid growth of plants by using artificial chemicals means they absorb more water, look bigger, but actually there's less food in them.

We're seeing the passing of these chemicals—non-biodegradable material—up the food chain, and we are storing toxins in our fat. So while breast is best, it's as good as the mother who makes it. So therefore it's not only essential for babies to eat organic, but it's also essential for pregnant women to eat organic, and to go organic well before conception—the sooner the better.

It's also vital for the father to eat organic prior to conception, as it takes three months to mature a sperm. The embryo and foetus are most at risk from chemical toxins and conception is the most critical time, followed by the first month, the first three months, the pregnancy and the entire infancy. Going organic will make a significant difference in cutting down you and your baby's total chemical load and giving your baby the best start in life."

Leslie Kenton (bestselling author and leading health expert, and mother of four):

"Organic babies get a head start in life simply because they have more of what they need to live at the highest levels of wellbeing, to develop fully and to remain resistant to degeneration as they move into adulthood. The organic matter in healthy soil is Nature's factory for biological activity. It is built up as a result of the breakdown of vegetable and animal matter by the soil's natural 'residents'—worms, bacteria and other useful micro-organisms. The presence of these creatures in the right quantity and type gives rise to physical, chemical and biological properties that create fertility in our soils and make plants grown on them highly resistant to disease. When it comes to human health they do a lot more.

The minerals and trace elements we need to trigger our metabolic processes on which health depends need to be in an organic form—that is, taken from living things like plant or animal foods. It is the organic matter in soils that enables plants grown on them to transform inorganic iron and silica into the organic form which is taken up by the vegetables and fruits, grains and legumes grown on them which we then eat, making these nutrients available to our bodies. Destroy the soil's organic matter through chemical farming, and slowly but inexorably you destroy the health of people and animals living on foods grown on it. Organic methods of farming help protect against significant distortions in mineral balances—an increase in one or more mineral elements can alter the availability of others. This can undermine people's health. No such protection is available when foods are chemically grown.

As a result of many years of eating nutritionally depleted foods, multiple deficiencies have become widespread. According to large scale studies, few people in the West still receive all the minerals they need to ensure that the metabolic processes work adequately—processes on which health and the immune system depend. Our indiscriminate use of chemical pesticides, herbicides, insecticides and other chemicals has quite literally poisoned the land we live in and contributed to widespread degenerative diseases. In most countries it grows worse year by year. Like ostriches, we have tended to bury our heads in the sand and hope that what we don't see won't hurt us. Meanwhile each year more than one billion gallons of chemicals are sprayed on to crops.

Raising your baby organically helps enormously to protect his or her whole life from beginning to end—especially if as an adult he or she continues to live on organic foods. I personally would not raise a baby any other way."

Dr Marilyn Glenville (author and mother of three):

"There has been a decrease in sperm counts by 50% over the last ten years, and an increase in testicular cancer. We now have a situation where girls seem to be entering puberty earlier than they did in past generations. At the turn of the century the average age was fifteen. Now some girls as young as eight are growing breasts and pubic hair. It has also been found that girls can enter puberty almost a year earlier if their pregnant mothers are exposed to higher levels of some synthetic chemicals while they are pregnant. These changes are frightening and are affecting the most crucial aspect of life—reproduction. It is now known that pesticide residues and other xenoestrogens (foreign oestrogens) are affecting fertility in the wild by altering the sexual characteristics of animals. Fish have been found which are hermaphrodite, having both male and female sexual organs It would be arrogant for us to think that as humans we are unaffected by these chemicals. In my practice, I specialize in the nutritional approach to fertility and female hormone problems, and it is apparent that fertility is declining, with one in six couples having problems. And yet, by asking couples to make simple lifestyle changes, like eating organic foods, eliminating alcohol and smoking, correcting vitamin and mineral deficiencies, etc, their fertility can often be restored. The dramatic effect of eating organic foods was illustrated in an article in *The Lancet* in 1994 which showed that organic farmers who ate organic vegetables—without pesticides and chemical fertilizers—had almost double the sperm count of men from other professions such as engineers and electricians who didn't. It is important that a healthy lifestyle including eating organically, is put into place before conception so that the egg and sperm are as healthy as possible. This also means that while the baby is developing in the womb, the number of chemicals it is exposed to are minimal. It is suspected that these xenoestrogens can affect the growing baby—the number of boys born with undescended testes has doubled since the 1940s. An article in *The Lancet* in 1993 suggested that exposure to these oestrogen-like substances during pregnancy could result in reproductive problems in the male baby later on, including the possibility of setting the stage for reduced sperm counts. By eating organically, we are helping to reduce not only the negative effects of pesticides on our fertility, but also on the reproductive health of wildlife. We are also increasing the chances of having a healthy baby and protecting future generations from fertility problems."

Michael van Straten, leading complementary health expert,
author of *Organic Superfoods* (from which this text is extracted)
and *Super Juice*, and father of two:

"There will never be a more critical period in your child's life in which to avoid exposure to damaging substances than during pregnancy. Even if the levels of chemical residue in non-organic foods are very small, if they get into you, they will get into your developing baby—and minute doses of toxic chemicals could be potentially devastating to a developing foetus. Crop residues are not the only problem. Illegal growth hormones have been discovered in meat purchased from butchers' shops throughout the EU. Excessive levels of antibiotics, found in dairy products, may cause the development of resistant strains of bacteria in babies and young children. None of these problems occur in organic food. What's even more reassuring is that not a single case of mad cow disease has been documented in cattle bred and raised on wholly organic farms. And without doubt, organically produced crops are richer in nutrients than their commercially grown counterparts. Non-organic produce tends to have a much higher water content, and thus contains fewer essential vitamins and minerals—which is why studies show consistently higher levels of vitamins A and C in organic fruits and vegetables.

Do you need more reasons for starting yourself, your partner and your developing child on an organic path? That path begins long before the baby does—three months before conception, in fact, when prospective mothers and ideally fathers should be feeding themselves organically. This gives your baby the greatest chance of growing and developing healthily in the womb. By doing so, he or she can come into this world in the best possible condition to survive, thrive and enjoy a healthy childhood.

Chemical residues in foods also pose a threat to the developing infant. Your baby's central nervous system and immune system are immature, making it more vulnerable to damage from toxic residues. In the UK, government statistics reveal that some British-made babyfoods contain many times the safe level of pesticide residues allowed by EU law—and a significant percentage of them harbour more than one. Some leading allergy experts believe that it isn't atmospheric pollution that causes asthma in children, but possible chemical damage to their immune systems. Add to this the huge list of artificial additives that find their way into various foods and you'll see why it is vital to use as much organic produce as possible when feeding babies and children."

Chapter 1

Why go organic for a healthy baby?

Instead of being nourished by their daily food, our babies—who are especially vulnerable to toxins—stand a very good chance of being poisoned by it. It is estimated that if you eat an non-organic diet, you consume an average of thirty pesticide residues (around 150 microgrammes of pesticides) every day. Much of our fresh fruit and vegetables contains detectable residues: in 2000, 72% of supermarket apples and 71% of pears. So every day we're unwittingly feeding our families a man-made chemical cocktail on the surface and in the flesh of our fruits, vegetables, nuts, beans and grains. Most of these pesticides can't be peeled or washed away (and most won't be destroyed by cooking). These chemicals are stored in our body fat and concentrated in our breastmilk, ready to be handed on in increased toxic concentrations to our precious babies. We're eating residues of antibiotics, hormones

The following list shows the incidence of residues in supermarket fruit and vegetables according to the last three annual reports published by the Pesticide Residues Committee (PRC).

Pesticide residues in supermarket food 1998-2001

Supermarket	% fruit and veg with residues	number of samples
Somerfield	60	109
Safeway	49	349
Marks & Spencer	49	94
Tesco	48	716
Sainsbury's	48	587
Asda	48	421
Waitrose	46	120
Co-op	43	109
Morrisons	42	173
Total	48	2678

(Friends of the Earth analysed data in the Annual Report of the Working Party on Pesticide Residues 1998, 1999, 2000, 2001 Brand Name Annex)

Pesticides in our food: government figures

On 31st May 2001 the Government's Pesticide Residues Committee published its quarterly monitoring report, for July to September 2000. The results make depressing reading for anyone who wants to buy food free of risky chemicals. They include:

• Cucumbers: 26% of Spanish cucumbers contained residues.

• Lettuces: 56% of samples contained residues, with multiple residues in 42%. Residues exceeding legal limits were found in 5 UK samples. Iprodione, a suspected hormone disrupter, was found at levels exceeding legal limits.

• Plums: 23% contained residues.

• Pears: 62% of samples contained residues. Only 3 UK samples contained residues but only 7 UK samples were taken compared to 75 imported. Carbendazim—a suspected hormone disrupter—was one of the most commonly found pesticides. Chlormequat was found commonly on imported pears and in one UK sample—although its use is not allowed in the UK.

• Milk and bacon were free of residues, but most of the oily fish contained residues including DDT and dieldrin which are subject to widespread bans because of their health effects.

All the major supermarkets were found to be selling food with pesticide residues. Lettuce samples with levels of pesticides over legal limits were found in Asda, Co-op, Safeway, Somerfield and Morrisons stores. A sample of UK pears from Sainsburys was found to contain chlormequat —a pesticide not approved for use here. Friends of the Earth (FOE) wants all retailers to aim for zero residues in their food as well as prohibiting the use of the most dangerous pesticides including those with strong evidence of hormone disrupting effects.

FOE also wants the Government to take action to ensure that our food is pesticide free including the introduction of a pesticide tax and better support for farmers to convert to organic or to reduce their use of chemical inputs.

Commenting on the results, FOE Food Campaigner Sandra Bell said:

"Yet again dodgy pesticides have been found in our food—from our own farmers and abroad. It's not surprising that consumers are clamouring for organic food, which we have to import in large quantities. The Government must make sure that farmers in the UK get the support they need to produce pesticide free food—and it should increase its vigilance on imports—why should consumers be expected to eat imported food containing pesticides which are banned in the UK? Labour's manifesto promises an investigation into sustainable farming and food—we will be watching closely to make sure that they deliver on it after the Election."

and other drugs routinely given to livestock and farmed fish (not to mention the chemicals, GM ingredients, and even the odd case of engine oil, slurry or dead stray dogs in livestock feeds), along with thousands of artificial chemical additives used in food processing. And as if that's not enough, new toxins are feared in food made with GM ingredients: 60–90% of processed foods are now thought to contain a genetically modified ingredient, or else genetic modification processes have been used in their production—most of which remains unlabelled, despite new EU legislation.

The government says there are no risks to human health from eating these (as it did during the BSE fiasco), but there are. Even very small amounts of individual chemical residues have been found to have profound health-damaging effects, and nobody knows the precise effect of eating multiple chemical residues (a cocktail of just two chemicals can hugely magnify their toxicity), nor the effects of eating GM foods. Over 60 pesticides, many of them used in the UK and on imported foods, have been named by the London-based Pesticide Action Network UK (formerly the Pesticides Trust) as causing adverse reproductive effects, and there is an ongoing campaign by six UK environmental groups to ban one of them: the organochlorine lindane, which is sprayed on UK fruits, vegetables and cereals, and believed to be a cause of breast and other cancers. The EU announced in 2000 that its use will be phased out in Europe. Eating intensively farmed food grown with such toxins is like playing Russian roulette—you never know when you, or your baby, are going to get a nasty surprise.

Experts around the world who have seen the continued rise in terminal cancers, infertility, reproductive disorders, declining sperm counts, birth defects, miscarriages, stillbirths, behavioural and developmental disorders, and diseases such as Parkinson's and Alzheimer's (all of which have been linked with the use of pesticides and other chemicals used in intensive farming), can only fear the worst for our children. There is growing scientific evidence that many diseases and susceptibility to illness in later life stem from exposure to toxins and poor quality food in childhood.

So what is safe to eat, who can we listen to, and who can we trust? And most importantly, just what is safe to feed our babies?

ORGANIC (INCLUDING BIODYNAMIC*) WHOLEFOODS offer the answer: they are the safest, healthiest, most natural, and arguably the most nutritious foods you can give your baby. They are grown without toxic artificial pesticides, genetically modified organisms (GMOs), and the many other routinely used agricultural inputs and practices which spawn the regular food scares. Organic groceries are also free from the thousands of unnecessary chemical additives used in conventional food processing, many of which have been found to cause or aggravate allergic reactions and contribute to hyperactivity in children.

In several studies in the UK and US, organic foods have also been found to have more vitamins and beneficial trace elements than chemically treated foods. With organic food now readily available and becoming increasingly affordable, you'll know exactly what you're eating and serving up to your family at mealtimes, and where it comes from. More details on the many benefits of going organic are given below.

Twenty-two good reasons to go organic for a healthy baby

• **You can trust the food**—certified organic food is grown and (minimally) processed to strict national standards which are routinely checked, and the food is easily traceable. Only high quality farm inputs and ingredients are used.

• **You can trust that what's on the label tells the full story.** Processed organic foods are fully labelled with the exact details of all ingredients used, unlike many regular brands which can contain 'flavourings' and other non-specific extras such as 'extracts', or can even leave off the label altogether some artificial additives used in processing, which can remain in the food; and they can leave out other ingredients, if they make up less than 5% of the product. A 'GM-free' label on an organic product legally means just that: no GM material is permitted (although there is always the possibility of inadvertent contamination); whereas a non-organic product labelled 'GM-free' can in fact contain up to 2% GM ingredients. New EU legislation approved in July 2002 will reduce this to 0.5%. Organic foods—governed by strict labelling regulations—are also clearly marked with the reassuring organic symbol of their certifying organic body, such as the Soil Association or Demeter, and may only be described as 'organic' on the label if they contain 95% or more certified organic ingredients. One organic babyfood brand, Baby Organix, has led the way with full percentage listing of ingredients.

• **You're avoiding or significantly reducing your baby's exposure to harmful pesticide and other man-made chemical residues.** Organic food is grown without artificial chemicals such as fertilizers, pesticides,

---

* Biodynamic agriculture is based on the principles of the Austrian philosopher Rudolf Steiner. Planting and other farm activities follow a calendar based on the movement of the moon, the planets and the stars, and involve the use of special natural soil and plant preparations. The certifying body, Demeter, is an internationally approved organisation, like the Soil Association.

fungicides, sprout inhibitors, drugs, growth retardants and antibiotics used in intensive farming, residues of which are found on and in our food (see box on page 18). In 2001 alone, over 26 million kilograms of pesticides were sold in the UK, and most of this was sprayed on to Britain's fields. Pesticide residues in amounts many times over the 'acceptable' safety limit have been found in government samples of fruit and vegetables. Infants are more at risk than adults from these residues, due to their high food intake compared with body size, restricted range of foods, and fragile and undeveloped organs and bodily systems.

There is mounting scientific evidence that many diseases and susceptibility to illness in later life stem from exposure to toxins and poor quality food in childhood. Pesticides and other factory farming chemicals can cause degenerative diseases such as cancer, hormone-related diseases such as breast and prostate cancer, nervous system disorders such as Alzheimer's and Parkinson's diseases, reproductive disorders, birth defects and even deaths. Some are known to mimic hormones or be hormone disrupters or blockers. Pesticide residues, including 87 dioxins, have been found in over 350 man-made chemicals present in human breastmilk (see box on page 29), and the breastmilk of mothers in the UK contains between ten and forty times the maximum

"We both ate organic food before we conceived, and used herbs to strengthen our immune systems and help our reproduction, and took supplements for magnesium and calcium before and during pregnancy. Our twins are now five and a half years old and eat mainly unprocessed fresh organic food and have never been to a dentist—we're strict about them not having sugar. We give them foods as close to nature as possible, but have found some organic products such as organic apple juice handy. They have only had very slight colds while other people's children seem to get ill regularly—especially with ear infections and gastroenteritis—in fact many other families seem to come down with this and other illnesses all the time. They're not vaccinated, and were breastfed. The only time they have been sick was when we went travelling to India—when they weren't eating organic food—and they got measles and had high temperatures for a few days, but we gave them echinacea and my husband did reflexology on them. We don't use sunscreen on them but they have never got burnt and we don't shampoo their hair —their hair stays clean, we wet it occasionally in the bath, but it cleans itself and the twins both have beautiful blond hair which people often remark on. They sleep well—they slept with us in our big futon bed as babies, which was very handy for breastfeeding, and now they have their own single bed in the same room, which is nice, and we have spare rooms! As twins they have been comparatively easy, which I believe is a lot to do with diet."—Mary-Clare Buckle, Abbotsbury, Dorset.

amounts as specified by the World Health Organisation's. Consuming low levels of antibiotics in food over a period of time can interfere with and suppress a child's immune system, making him more susceptible to colds and infections, and is also thought to be a contributor to asthma. Potentially life-saving antibiotics, when needed, can be ineffective.

• **You're protecting your baby against the unknown health effects and potential toxicity of genetically engineered foods.** Some of these foods have been shown in preliminary UK and US studies to deform and kill butterflies, adversely affect other useful insects such as ladybirds, and impair the immune system and alter the growth of internal organs in laboratory rats. Organic food is certified to be GM-free, and organic organisations are actively campaigning against the introduction of commercial GM crops, which are likely to lead to increased pesticide usage through the development of plant and insect resistance; higher pesticide residues; and the contamination of organic crops by cross-pollination. Campaigners fear that organic food production will no longer be possible in this country if commercial GM food production goes ahead (see the section on GM concerns on page 31).

• **You're protecting your baby from the thousands of artificial additives and E numbers** such as chemical preservatives, many hidden sugars, fillers, emulsifiers and artificial sweeteners (some of which have been linked to cancer), flavourings and colourings used in conventional food processing. In 2002, a UK government-sponsored scientific study has corroborated (for the first time) the link between food additives and changes in children's mood and behaviour. Research scientists at the UK's Asthma & Allergy Research Centre, working on behalf of the Food Standards Agency, concluded that 'significant changes in children's hyperactive behaviour could be produced by the removal of colourings and additives from their diet.' Many organic foods are also sugar- and salt-free, or contain much less of these ingredients than regular brands, and don't contain health-damaging hydrogenated oils or fats (see box on page 86).

• **You and your baby are supporting a healthier farming system** which ensures a high standard of animal welfare and which produces high quality, uncontaminated meats and dairy foods. Organically reared animals are not routinely kept in confined spaces, abused or fed drugs— they roam free on grassland and have a varied and largely organic diet. Animals are not fed other animals (whereas conventional cattle are still able to be fed poultry parts, and poultry can be fed fish) and no cases of BSE have ever been recorded on an organic farm. Nor do they have artificial colourants added to their food to produce yellow egg yolks (as do battery hens) or to make their flesh pink (as do farmed fish).

• **You and your baby are encouraging a more stringently controlled food production system** which has a high standard of hygiene, and is much less likely to be susceptible to the hallmarks of mass produced food: E. coli, salmonella, listeria and other harmful bacteria which cause millions of cases of food poisoning in the UK each year, and annually cost the taxpayer billions of pounds.

• **Your baby's food will have increased vitamins and minerals, trace elements and other important nutrients** compared with non-organic foods. Several UK and US research studies have shown organic foods to have many more vitamins and minerals, and more vitality than their conventional equivalents. Chemically treated food is forced to grow faster and bigger with artificial fertilizers, and nutrients are often lost in the process. Deficiencies of key nutrients in our diets, such as selenium, chromium and zinc, are increasingly being found; they can make children susceptible to illness and diseases such as diabetes, infertility, and even cancer. Research has shown that IQ level in babies and children improves with increased vitamins and minerals.

• **You're safeguarding your baby's fertility in the future.** Organic food has been found to increase fertility, probably owing to its increased nutrients such as zinc, and the fact that it doesn't contain pesticide residues. Danish researchers found that organic farmers and men who regularly consumed organic food had twice the sperm count of men who did not. In the UK, the Foresight Association for the Promotion of Pre-conceptual Care has shown that where there were previous infertility problems, eating an organic

"We've been eating organic for a few years, and certainly long before I had my son, who is now two. I don't know whether it had anything to do with it, but the pregnancy and birth went really well, and certainly my son has been very healthy ever since—he's only ever had a couple of colds, while other parents are always telling me about their children being sick and having ear infections, etc. He's bigger and sturdier, and much healthier looking than other children around us, and he seems to eat more as well—he's not fussy at all. I don't give him crisps or any sweets as other parents are always giving theirs and so he's developed really good eating habits. He's always slept well, and part of that may be because he slept in our bed as a baby and is still in our room, in a bed of his own. I wouldn't have had it any other way—it just felt really natural and lovely having him there with us and it would feel really strange for him not to be. I guess he'll decide when he's ready to have a room of his own. And I still give him one breastfeed a day. My husband's allergies have also been much better since we went organic—it's really made a difference to his health."—Helen Stafford, St Johns, Exeter.

wholefood diet can help to correct these problems and enable conception to take place without the need for expensive IVF treatment and drugs.

• **Your baby will enjoy his food more,** as organic food tastes better. Most taste tests of conventional versus organic food show a marked preference for organic, which is partly why leading chefs such as Antony Worrall Thompson, Raymond Le Blanc and Jamie Oliver have become converts.

• **You're contributing daily to your baby's own personal healthcare scheme** by building a foundation of health, and a strong immune system which will help to protect him against illness, rather than burdening his system and programming him for disease later in life.

• **You and your baby are increasing the demand for organic food,** sending a strong message to retailers, food manufacturers, farmers and the government that you don't want to buy intensively farmed food. This means that more farmers will grow organic food and more retailers will sell it, increasing the number of retail outlets and organic food delivery schemes—making organic food more easily available. It will become more affordable, available in larger quantities, and in increasing varieties and quality. The government will be forced to give greater support to the industry, diverting funds which would otherwise have supported conventional agriculture.

• **You and your baby are supporting sustainable farming,** which recycles natural resources rather than depleting them. Unlike intensive farming, which strips the soil of nutrients and applies artificial ones, organic production builds the health and fertility of the soil with composts, green manures and cover crops, and crop rotation. It is then able to produce healthy and strong plants and animals, resistant to disease. While pests are found in organic crops, their natural predators are encouraged, and natural systems are in balance. It surely follows that with healthy soils and healthy plants and animals, there will be healthier humans at the top of the food chain.

• **You and your baby are reducing the pollution caused by intensive farming,** which contaminates groundwater, rivers, streams and lakes with the over-use of nitrates, pesticides and drugs used to treat farmed fish. Following a 1996 government report which revealed that more than 21,000 UK drinking water samples contained pesticides over legally 'acceptable' concentrations, new, more stringent regulations were introduced in 2000. Although the quality of drinking water is improving, the cost of extracting pesticides and other pollutants is expensive—and a cost borne by the consumer.

- **You and your baby are helping to protect and encourage wildlife** such as earthworms, insects, birds, bats, badgers, fish, otters, voles (which are nearing extinction) and butterflies, and the wildflowers, trees, waterways, and habitats they depend on. These habitats have been progressively destroyed and polluted by intensive farming, and several UK species are already extinct or have been endangered as a result. Male fish in UK streams have developed female sex organs; this is believed to be caused by artificial chemicals which mimic oestrogens. Organic farmers encourage wildlife to flourish on their land and promote biodiversity—birds and butterflies have been shown by the Royal Society for the Protection of Birds to occur in greater number on organic farms. They are considered an essential component of a balanced ecosystem.

- **You and your baby are helping to preserve the countryside we all** love—our cultural heritage—with its unique rural landscape features such as small fields, traditional orchards, ancient meadows and hedgerows, which are most commonly destroyed by intensive farming, yet are prized by organic farmers. Some areas of England have become prairie monocultures due to intensive farming practices by cereal growers, and it is estimated that up to 10,000 miles of hedgerows are lost in Britain each year. Many traditional fruits, vegetables and farm animals have also disappeared as intensive farming favours only a few varieties.

- **You and your baby are helping to protect the ozone layer and save energy**—chemicals such as methyl bromide, which is sprayed on conventional strawberry crops, damage the ozone layer and contribute to global warming, and require large amounts of non-renewable energy to produce. Methyl bromide is considered to be 60 times more damaging to the ozone layer than the banned CFC chemicals.

"My daughter had rashes from allergies when she was just a few months old, and when we had her tested the practitioner said that her allergies to tomatoes, potatoes and dairy produce were due to the hormones given to the cows and the chemicals in the food. We changed to organic and now she can eat those foods. She also used to get terrible tummy pains after eating meat, but with organic meat her digestion is much better and there are no pains. It has made such a difference as her diet is no longer so restricted and there's a remarkable difference in her skin. I now believe that the only way to bring up a healthy child is by eating organic. I had a wheat allergy, but now I can eat organic wheat—and can have bread and biscuits again. Organic food is becoming more accessible all the time and I'm so glad that I can also find many organic alternatives to children's food, as the children like them for treats. "—Melissa Paulden, Berkshire.

• **You and your baby are supporting and encouraging ethical, environmentally friendly and sustainable businesses** (mostly small ones), practices and lifestyles. By their very nature, organic companies have a tendency to value people and the environment above profit.

• **You and your baby are helping to employ more people**, keep rural communities intact and small landholders on the land, at a time when the crisis in the conventional farming sector is doing the exact opposite. Organic farms hire many more people than conventional ones, as many tasks are done by hand rather than mechanically (for example, pulling out weeds). As the demand for UK-grown and processed organic food continues to increase, so too are job opportunities in the organic food industry.

• **You and your baby are helping to protect the health of farm workers** and their families, who aren't being exposed to dangerous chemicals. When buying products sourced from overseas, if you buy 'fair-traded' organic, you are also helping to secure a better deal for people in third world communities, who are paid a fair price for their crops, reducing the necessity to grow cash crops at the expense of staples and to use dangerous chemicals (many of which are banned in this country), which cause thousands of deaths and cases of poisoning each year. The community decides how the income from the sale of their produce should be used: they are building facilities such as schools, buying books and educating their children.

• **You and your baby are part of a 'true food' revolution** which is turning the tide on unsustainable, intensive farming and contaminated food, helping to make the world a safer, healthier and better place for us all.

• **Because you can!** High quality organic fresh foods, groceries, and ready-made babyfoods (see the Babyfoods section in Part Two) are now much more readily accessible and affordable than ever before. They're in healthfood stores, on supermarket shelves, in Boots, in a growing number of organic supermarkets and restaurants, and available from an ever increasing number of organic box schemes, home delivery companies and mail order services.

Many ways to make the most of these benefits of going organic— whether you're planning a baby, already pregnant, or already have a baby or older child—are outlined in Chapters Two and Three.

What price your child's health?

While babies are considered more at risk than adults from pesticides and other chemical residues and toxins found in our food, experts agree that they are even more at risk when growing in what was previously considered to be the safest place in the world for them—their mother's womb. The 8,000 babies severely deformed by the drug Thalidomide about forty-five years ago first demonstrated the ability of toxins to cross the placenta and harm the foetus, and today medical staff delivering the babies of mothers hooked on heroin or crack witness this when the babies are born addicted to the drug.

The sperm and egg, tiny embryo, and foetus are most at risk in that order— in the critical months before conception when the health of the sperm and egg is determined, when first developing inside the womb as an embryo with its rapid cell division, and then when growing tissue, bone, organs, brains and nerves, and rapidly developing until birth, as the foetus takes all its nutrients through the placenta from the mother's blood.

The Worldwide Fund for Nature (WWF) said in a 1999 report that more than 350 man-made pollutants have been identified in the breast milk of women in the UK, and that British babies fed on breast milk could be receiving as much as 40 times World Health Organisation recommended levels of a wide range of potentially harmful chemicals. These include 87 dioxins, the poison which sparked health scare in Belgium. The effects of these persistent, non-biodegradable chemicals on tiny foetuses and embryos in the womb is expected to be even worse— especially at key developmental stages. Dr Vyvyan Howard, a foetal toxicologist at Liverpool University, has been trying to raise awareness

"I believe our diets are responsible for a lot more than people give them credit—on a diet containing lots of raw organic fruit and veg we have more energy, think more positively, are more tolerant, and are happier in ourselves. It upsets me so much to see children being raised on a lot of processed rubbish—they are getting no nutritional goodness at all, and it really shows in their behaviour. My son Reuben is not a big eater, but I do not worry, because all he has is so packed with nutrients, I figure he doesn't need that much. Everyone says how bright he is, and I am sure that is partly due to diet. I try and buy organic wherever possible—my rule is as long as it is not more than twice as expensive as the non-organic product, I will buy it. I guess around 50 to 75% of the food we eat is organic. I think everyone knows organic is better, but people don't realize how bad non-organic is, especially for children—the risk of cancer or neurological disorders is so much greater, because they consume more in relation to their body weight."—Kate Sullivan, London.

of this increased risk to the unborn baby for many years. His work has found that if foetal development is disturbed by environmental toxins or the lack of key nutrients at key stages of growth, there can be severe retardation, such as only developing half the number of nephrons normally found in each kidney. London-based natural childbirth pioneer Dr Michel Odent claims that uterine pollution represents arguably the most serious threat that humanity has to face at the turn of the century.

So from conception onwards, unless you and your family eat organic, your unborn baby could be exposed to a toxic chemical cocktail derived from your food, with potential damaging effects. It's no accident that the incidences of miscarriages and stillbirths, cancers, malformations, disorders and allergies in our children (and also adults) are increasing at the same time as pesticide, antibiotic and other artificial chemical usage in our food production and processing continues to rise. The Pesticide Action Network UK has a library full of UK and international studies correlating these facts and showing detrimental reproductive effects of pesticides on both men and women, chemical causes of cancers, evidence of chemical residues in breastmilk, and evidence of reduced birthweight and length in the offspring of women exposed to pesticides, and related congenital malformations. The WWF-UK report *Chemical Trespass: a toxic legacy*, published in June 1999, also highlighted many of such studies. The Primal Health Research Centre in London has also created a databank of over 400 studies exploring the correlation between health during the 'primal period' (conception to one year) and severe health problems later in life.

Despite government assurances to the contrary, there is no safe level of these chemicals. For more than fifty years, warnings about this toxic chemical legacy, which is being passed on from generation to generation in ever increasing concentrations, have been made by scientists, organic and environmental pioneers and activists who foresaw the dangers, and who have seen the effects in both animals and humans around the world. This legacy now stretches to the furthest reaches of the earth, affecting eskimos, polar bears, seals, dolphins and whales alike, but pleas for official sanity have largely fallen on deaf ears. As we enter the new millennium we should help to turn the tide by refusing to continue to eat chemical food, by going organic and supporting organic farmers, manufacturers and retailers.

It may well be that a lot of modern day health tragedies—which are put down to misfortune and 'fate' by much of the medical community—can in fact be prevented by minimizing exposure to these artificial chemicals and by eating natural, wholesome organic foods during pregnancy and before babies are conceived.

While we're officially warned during pregnancy about drugs, alcohol,

Our toxic chemical legacy

More than 350 man-made toxic chemicals, many of them from residues in food, have been found in breastmilk and body fat, and are being passed on to foetuses in increased concentrations via the placenta—a toxic chemical legacy being handed on to the next generation. A WWF-UK research report *Chemical Trespass: a Toxic Legacy*, published in June 1999, documented the existence of many of these chemicals in adults and in breastmilk, and the fact that breastfed infants in the UK were ingesting through their mothers' milk up to 42 times the World Health Organisation Tolerable Daily Intake of dioxin-like compounds (as found in a WHO survey).

Chemicals found in UK breastmilk include toxic metals, dioxins, PCBs, DDT, the widely used organochlorine lindane, and other pesticides known to cause birth defects, infertility, trigger cancer, damage the immune system, cause developmental problems and disrupt sex hormones. The main source of dioxins and PCBs in the UK (according to MAFF) were milk and dairy products, fish and fish oils, and meat and animal fats.

The study reported that food, food packaging and consumer goods such as cosmetics, sunscreens and cleaning agents are the main causes of contamination, and excluded others which can also harm foetuses and babies via the placenta and breastmilk, such as tobacco, alcohol, cannabis, pharmaceutical drugs and nitrates.

"It is now recognized that the foetus can be damaged by relatively low levels of contaminants which do not affect the adult," stated the report. "Exposure in the womb can cause birth defects and affect our children's future ability to reproduce and their susceptibility to diseases, including cancer. Functional deficits can also be caused, such that some children may not reach their full potential. Put simple, the integrity of the next generation is at stake."

The study echoed the chilling findings and warnings published in *Our Stolen Future* by US scientist Theo Colborn in 1996, and by Rachel Carson in her ground-breaking book *Silent Spring*, published more than thirty years earlier, which first described the toxic chemical legacy being passed on to the next generation. Warnings of the dangers of intensive chemical agriculture, and ignorance of the 'healthy soils equals healthy people' principle were made in the 1940s by UK organic farming pioneers Lady Eve Balfour (who founded the Soil Association) and her contemporary Sir Albert Howard. UK foetal toxicologist Dr Vyvyan Howard says that we have between 300 and 500 synthetic industrial chemicals in our bodies that would not have been there fifty years ago.

cigarette smoking, and even toxoplasmosis (which can cross the placenta and harm your baby), pesticide residues and other serious byproducts of intensive farming in our food don't even rate a mention. Surely we should be encouraged by the government, and by leading organisations such as the National Childbirth Trust, to go organic so that we avoid these dangers? Such preventative health measures could save the NHS millions of pounds each year and potentially save many young lives and the good health of thousands of children as well.

This important role is instead left to a few 'alternative' health experts and the unsung heroes of small organisations such as the Foresight Association for the Promotion of Pre-conceptual Care, the renowned Surrey-based charity. Foresight, formed over twenty years ago, has the evidence to prove that many severe health problems in infants previously thought to be unpreventable can in fact be prevented by optimizing both parents' health before conception, thus maximizing the chances of a healthy sperm, egg and uterus. A healthy pregnancy drawing on a well-stocked health bank of pollution-free nutrients promotes the birth of a perfectly formed and healthy baby. Foresight's important work and excellent pre-conceptual programme are detailed in the next chapter, which also reviews the work of Dr Michel Odent and The Primal Health Research Centre in London. The Centre promotes pre-conception care, an organic diet and a revolutionary new way of reducing accumulated toxins in body fat before pregnancy and breastfeeding, called the Accordion Method.

AS ALL PARENTS want the best for their children, surely the best we can offer them is an organic foundation and future. Even though some may say that the cost of organic food is still too high to incorporate as the main component of a baby's or family's diet, others might ask: just what is a child's health worth? The cost of not eating a mainly organic diet is a price too high for our babies to pay.

With many new organic products coming on to the market each month, and with an increasing number of farmers and growers, manufacturers, retailers and specialist businesses supplying organic foods, goods and services, sheer economies of scale and competition are bringing organic prices down all the time. Some items are now the same price as regular ones, and some are even cheaper. Nevertheless, organic items generally remain more expensive, due to the true cost of producing real food (discussed later in this chapter). Useful tips on how to manage going organic on a limited family budget are featured in Chapter Three, and in the experiences of parents quoted in boxes throughout Part One of the book.

Whatever age your baby or child is, you can significantly improve his health, both now and in the future, by increasing the proportion of organic food he eats. Each and every change you make will be a positive step.

A few of the reasons why we should all be concerned about, and avoid, GM foods

According to several UK experts, developing foetuses, babies and children are likely to be most susceptible to any new toxins which are produced by genetically modified foods. These foods are created in the laboratory, are unnatural and unnecessary: they're products which consumers never asked for, and weren't given the choice as to whether or not they wanted to eat them. They are foods which consumers don't want (every consumer poll confirms this), and don't gain any benefit from, but are still being force-fed regardless. In July 2002 the European parliament approved sweeping plans for the labelling and traceability of genetically modified foods. These GM labels will apply to foods which contain more than 0.5% of GM material.

• 'GM-free' foods (non-organic) are permitted to contain up to 2% GM ingredients.

• Food containing genetically modified ingredients above the 1% threshold (soon to be 0.5%—see above) must have clear labelling to indicate that it contains GM ingredients.

• Food labelled as 'organic' may have no GM ingredients used intentionally.

There are grave concerns over potential health risks for people who eat foods with GM ingredients: the short and long term health effects of eating them are unknown, and will not be known for many years. Preliminary tests in the UK and US on laboratory rats and insects such as ladybirds and butterflies have shown severe detrimental health effects, and an early GM amino acid is thought to have produced a powerful toxin which reportedly killed at least thirty-seven people in the US, and permanently disabled up to 1500 others. Genes from nuts have been used in the development of new GM foods which could trigger potentially fatal allergic reactions in people already allergic to nuts, and the development of allergies in others.

Liverpool University foetal toxicologist Vyvyan Howard has publicly voiced fears for unborn babies, who would suffer the most from any toxins produced by GM foods. He believes the risks are not being properly considered. "Before Chernobyl, it was always considered to be a 'vanishingly small risk' that such a disaster could occur, until it did!", he says. GM foods and ingredients on the market have been unnaturally altered by scientists, who insert genes from different species which would never naturally be crossed (such as putting a fish gene into a tomato, which is of particular concern to vegetarians and vegans). They also

The dangers of GM technologies

Prof. Terje Traavik began by saying that he is a professional genetic engineer and has been for the past 20 years. At first, he was a total 'believer' in thinking that there were only benefits. But he changed his mind as the result of discoveries made in his own laboratory. "We have no gene technology!" he said categorically, basically because the so-called technology is uncontrollable and unpredictable, so much so that there is no basis at all for risk assessment. Perhaps the next generation of technology may deserve the label. He emphasized that the gene constructs are the same, and offer the same risks, whether they are used in agriculture or in medicine, such as gene therapy vectors and vaccines. Nature has never seen those sorts of genetic constructs before. They pose huge risks as they can become mixed up with normal viruses and other invasive elements and transfer their traits elsewhere. The potential hazards of artificial constructs are much greater than chemicals. Because, instead of breaking down or diluting out, they are taken up by cells to multiply, mutate and recombine indefinitely. It may be 'BSE [mad cow disease] in technicolor'.

"The foreign genes and constructs cannot be targeted and are inserted at random, causing all sorts of genetic disruptions known as epigenetic effects. These can give rise to the production of new toxins and allergens. All plants contain toxins and allergens but the toxins are produced at very low levels and are therefore not harmful. GM can result in the over production of toxins and allergens in the GM plants and present a serious health risk." He repeated his call for banning the first generation of GMOs.

From a Parliamentary Briefing to the UK Government in February 2000 by Professor Terje Traavik, senior scientific advisor to the Norwegian government. Courtesy the Institute of Science in Society (www.i-sis.org.uk).

insert plant viruses and antibiotic markers (which are believed to be potentially disastrous for public health as they may result in antibiotic 'superbugs'), the aim being to create a new man-made life form, most of which are designed to be resistant to herbicides. There is no evidence of increased nutritional value or other benefit to the consumer: these foods have been developed for the financial benefit of the herbicide-selling companies who created them, in some cases to prolong the shelf life of the product, or to make them better able to withstand handling and transporting. Genetic modification is expected to increase the amount of herbicides used (higher pesticide residue levels found in GM soya have led the government to raise the acceptable limit of the pesticide glyphosate (Monsanto's Roundup) by 200 times), to cause greater destruction of wildlife and biodiversity, to increase livestock suffering and to create 'superweeds' resistant to herbicides. GM crops are also expected to contaminate native plant species, farmland and honey—and once released, they can never be removed from the environment.

According to London-based GE expert Dr Michael Antoniou, the GE process of randomly splicing a foreign gene into the host DNA always disrupts natural genetic order and function, giving GM foods an inherent unpredictable component that is greater than the intended change. "Disturbance(s) in genetic function can lead to disruptions in the biochemistry of an organism which can result in the production of novel toxins and allergies," he says. "Criteria used by regulatory authorities to assess health risks of GM foods in the UK and worldwide, do not take into account the unpredictable side to GM technology which basic genetic science tells us is present. As a result foods are released on to the market without adequate testing for potential allergies and toxic effects. As with any potential toxic and allergic effects from any source, problems arising from GM foods will most likely be highlighted in the young in the first instance due to their greater sensitivity."

As well as many leading environmental organisations, national figures and institutions, several leading supporters of GM technology have recently publicly raised serious misgivings about the risks of GM foods, and many UK farmers are now refusing to carry out trials on their land because of growing public outcry, fears of human health risks, contaminated land and subsequent reduced land value. Meanwhile, the Blair government continues to support GM food technology and development and thousands more new GM foods are in the pipeline.

As public resistance to GM foods has gathered momentum, the biotech companies have begun to develop a 'second generation' of GM crops, including so-called 'functional foods', which claim to give significant health benefits. Functional foods already on the market include examples where the product has vitamins added, or where an allegedly more healthy form of fat is used in the food manufacturing process. Biotechnology is increasingly being applied to produce functional foods: the industry claims that humanity will benefit if vitamins are bred into basic crop staples in Third World countries to stave off vitamin deficiencies, especially in children. Other promised novelties include tomatoes that contain 'cancer-fighting substances', allergen-free peanuts, and for vaccines to be bred into fruits like bananas to tackle common diseases like Hepatitis B.

The only way consumers can now guarantee that they are not eating GM foods is to eat organic, which is certified GM-free with a zero tolerance level of GM ingredients. But even organic food and farming is at threat from this new technology due to contamination by cross-pollination (as bees and pollen travel many miles), and campaigners believe that unless crop trials are stopped and the commercial introduction of crops is banned, there will be no organic option in the future. Many of the GM field trial sites in the UK have been targeted by protesters, and public resistance has been such that the Government has agreed to ask the public what its views are before it makes a decision on whether to allow GM crops to be grown commercially in the UK. The report by the government's Chief Scientific Adviser, Professor David King, is due to be completed by June 2003.

It seems unbelievable and illogical that the UK government is still supporting the introduction of potentially dangerous foods which consumers are consistently saying they don't want and which will contaminate the organic foods they are demanding in ever-increasing numbers—foods which are being produced by the only thriving sector of the entire farming industry, and foods which are employing an increasing number of people and helping to rebuild a depopulated rural society, and polluted countryside.

There are now some excellent books on GM foods which are recommended reading (see Publications in Part Two); you can also consider supporting organisations that are campaigning against GM foods on your behalf such as Greenpeace, Friends of the Earth, the Genetic Engineering Network and the Soil Association (see Organisations in Part Two). As a consumer, you can protest against the introduction of these foods whenever you shop: boycott any brands made with GM ingredients, or which can't give you a guarantee that they're not, and of course always choose organic if possible.

Why organic is generally more expensive —and why cheap food isn't cheap at all

While organic foods are becoming more affordable, they are still generally more expensive because their price reflects the true cost of their production. Organic farming is more labour-intensive, is largely unsubsidized, conserves and manages traditional landscape features (not only if there are cash incentives), uses high quality livestock feeds, must meet strictly enforced higher standards, and does not take short cuts for the sake of profits. (Also see other benefits in the 'Twenty-two good reasons to go organic' on pages 20-27).

Unfortunately we live in a world where polluting ('grow fast, big and in large amounts at any cost') intensive agriculture has billions of pounds thrown at it by the government and EU each year through subsidies, research assistance, grants, direct financial support and cash incentives—

money from our taxes, which could be used for far more deserving causes such as further research into organic agriculture, increased assistance for farms converting to organic, and preventative health education in the community. Unsustainable intensive farming employs fewer people each year, contributes increasingly smaller amounts to local rural economies, as big business continues to buy up and amalgamate small farms, sending the money elsewhere, and actually puts an increasing number of smaller farmers out of business (the industry is currently in the midst of its biggest ever crisis, while organic farming and farmers are thriving). It is a major polluter and consumer of energy and natural resources, which it depletes, and it turns the traditional rural landscape into prairie monocultures in its drive for ever greater profits.

According to *The True Cost of Food*, a report from Greenpeace and the Soil Association, if elements such as air and water pollution, eroded soils and health care costs were factored into the price of produce, organic produce would be the same price or even cheaper than food produced conventionally. We're actually paying several times over for 'cheap', potentially poisonous and unsafe food (through our taxes and from our wallets when we buy the food), and paying even more if we want the 'privilege' of eating naturally produced, healthy, safe, unadulterated food, which should in fact be everyone's birthright. We also pay with our own and our children's health as we eat contaminated, nutrient-lacking food, and through our water bills as we're charged for some of the millions of pounds spent each year by water companies to clear drinking water of farm pollutants such as pesticides and nitrates—many of which nevertheless remain, further damaging our health. The cost to the environment is even higher (and often irreversible), and we pay again to support worthwhile charities to campaign against the pollution and contamination, to try to save endangered wildlife, and to support cancer and other disease victims, research and develop 'cures', and promote health prevention. So cheap food isn't cheap at all: it actually costs us all dearly, while organic food has no hidden costs and in fact saves society a great deal of money and provides numerous valuable benefits to communities, public health and the environment.

Consider the thought that through our taxes each year we're all unwittingly:

• **subsidizing the use of poisonous artificial pesticides and fertilizers** (which aren't taxed) and directly supporting—even rewarding—damaging and sometimes absurd farming practices which endanger public health, and destroy and pollute the environment.

• **paying to clean up the mess of intensive farming fiascos** such as BSE (£4.6 billion up to April 2002—over £200 per household);

• **paying to detect residues of harmful substances** which shouldn't be put on or in our food in the first instance (about £2 million a year is spent by the government's Working Party on Pesticide Residues to test for pesticide residues), and to monitor and regulate pesticide usage on farms and to deal with and prosecute abusers;

• **paying to support out-of-work conventional farmers** and their families damaged by pesticide poisoning and those forced out by the current farming crisis;

• **paying to administer a food standards agency,** and yet more committees to monitor and regulate conventional food processing (when it's the production system which is the real problem);

• **paying to debate, regulate and enforce regulations,** and manage and monitor worrying developments such as GM foods (which the public don't even want);

• **paying to fund an agriculture ministry** and ministers who manage, promote and preside over all of the above, and fund additional working parties to review, and try to reduce the amount of, 'red tape';

• **paying to support an ailing health service** treating an ever-increasing number of sick people at the end of the polluted food chain (there are around 4.5 million bouts of food poisoning in the UK every year, costing the taxpayer an estimated £350 million in hospital care and time off work).

In September 1999, the government also announced a £150 million aid package (of taxpayers' money) to conventional farmers hit by the current crisis, which, along with EU aid (also from taxpayers) will give them an additional amount of around £500 million. As you can see, it really doesn't make any sense for us—or for the government—to support the unsustainable production of cheap food with billions of pounds of public money each year, and conversely it makes incredibly good sense to support the production of organic food. So next time you look at your weekly food bill or the shelf price of an organic food compared to the price of a 'cheaper' food, and when you see your monthly or yearly tax bill, think about what it is that you're actually paying for. Is it worth it? And remember that the more consumers demand and buy organic food, the more will be grown and sold in this country, and at more affordable prices (around 65% of organic food is currently imported, but thanks to increasing demand, more land is now in conversion to organic than there is already being farmed organically, and major supermarkets are publicly competing to source, develop and stock more organic products). People power alone will ultimately determine the food we eat, the government which presides over us, and what our taxes are spent on—so cast your vote at your local retail outlet today! (For a full discussion of the politics of farming, read Graham Harvey's *The Killing of the Countryside*, published by Jonathan Cape.)

More than just a food issue

Organic no longer means just food. As well as a growing abundance of quality organic fresh foods and groceries, including those tailor-made for weaning infants (see Babyfoods in Part Two), there's now a diverse range of organic bodycare, nappies, clothing, bedding, herbal remedies and other useful products available by mail order and in healthfood stores for babies, pregnant mothers and the rest of the family. So if you're going to eat organic, you could consider taking the next obvious step: replace some or most of your consumer goods with organic ones, or at least choose environmentally benign ones, if no organic option is available. Residues of chemical toxins found in everyday consumer goods such as cosmetics, sunscreens, bodycare toiletries and cleaning agents have been found in significant amounts in breastmilk, alongside the pesticide residues and other contaminants from food, and are part of the chemical cocktail to which developing embryos and foetuses in the womb are exposed.

Unfortunately, much of today's clothing, soft furnishings and bedding contain unwanted additives which can add to our chemical load and cause or aggravate conditions such as eczema and asthma. Many materials are treated with flame retardant and moth-proofing chemicals, some of which have been associated with cot death by UK and New Zealand researchers. Formaldehyde and other chemicals can be found in 'easy care' and 'non-iron' synthetic fabrics, and may remain in these textiles for life. Strong bleaches and dyes (which can contain heavy metals, create cancer-causing dioxins and pollute waterways) are also used, along with other artificial chemical treatments, such as fungicides. Cotton is one of the world's most sprayed crops (using about a quarter of the world's insecticides, and causing many health and environmental problems, especially in third world communities), and residues may remain in the fabric for a long time.

Formaldehyde is also used in bed and mattress production, along with solvents and foams. According to the Breakspear Hospital in Hemel Hempstead, renowned for its allergy and environmental medical practices, as many as one in five people may have a sensitivity to formaldehyde, which it says is the most widespread chemical produced by petrochemical products such as plastics, textiles, paints and glues. It can cause symptoms in concentrations as little as one part in a million—concentrations often reached inside modern homes.

Disposable nappies are an environmental waste disaster, with an estimated eight million used and thrown away in the UK every day, adding to the landfill problem and contaminating groundwater. There are also grave concerns about the effect of the absorbent gel, dioxins and plastics used in the nappies themselves on baby's developing organs (see Nappies in Part Two).

Common cleaning agents contain many unlabelled chemicals such as hydrochloric acid, chlorine-based bleaches and petrochemical-based solvents that can cause mysterious rashes, headaches and allergies. These are harmful to the user, to the waterways they end up in, and to the dependent wildlife (not to mention our drinking water, which is drawn from these waterways). Babies and toddlers explore their surroundings with their mouths, licking floors and surfaces, and trying to drink from or suck any containers they find; in 1999 over 30,000 children under five were hospitalized after suspected poisoning by common domestic substances. These are very good reasons to keep them out of your home. Many bodycare toiletries contain a myriad of artificial and sometimes harmful ingredients (see Part Two for details), many of which are absorbed through the skin, and some of which pollute waterways and damage wildlife. Hair sprays, fly sprays, air fresheners, oven cleaners and many other common household products contain chemical ingredients which are harmful to your baby, even if he doesn't come into direct contact with them, and can cross the placenta while the baby is in the womb. Many pregnant women will find that they are naturally repulsed by the smell of such products anyway—your unborn baby's way of making sure you avoid them. Details of the dangers and problems caused by all of these products are given in the relevant sections of the Products Guide in Part Two.

Thankfully, there are now many (increasingly affordable) alternatives to these everyday products. There are organic cotton and woollen reusable nappies (cheaper in the long run—and some local councils are now subsidizing use of reusables to reduce their landfill cost); baby, children's and adult clothes; blankets, mattresses and even toys. You can also buy organic food supplements, soaps, shampoos and toothpaste, and choose from many environmentally friendly cleaning agents, which are reviewed in Part Two. Your baby can finally eat, drink, sleep, play, wear and bathe organic! So when you're going organic, think of more than just food.

Chapter 2

Healthy parents equals healthy offspring
Organic pre-conception, pregnancy & breastfeeding

Thousands of UK couples who have suffered miscarriages, stillbirths, infertility or given birth to malformed babies, try the Foresight programme, which involves converting to a completely organic, wholefood diet many months before trying to conceive a child. Almost 80% of all couples who undertake the programme conceive, and go on to have healthy and perfectly formed babies (whereas the average success rate for IVF is 22.7%).

Pre-conceptual care
Investing in parents' and baby's organic health bank

MANY EXPERTS NOW AGREE that we ignore at our children's peril the importance of creating healthy sperm and eggs, a pollution-free womb, and a sufficient health bank of all the nutrients needed for a rapidly developing embryo and foetus.

It's the organic farming principle of growing healthy food from healthy soil, applied instead to creating a healthy baby. Organic farmers pay great attention to building up the long-term health and fertility of their soil—studiously preparing the ground with natural, recycled, biodegradable nutrients over time, so it will be enriched and contain both the nutrient mix and micro-organisms needed to fully nourish the seed and support its growth. As a result they produce strong, vigorous healthy plants that are naturally resistant to diseases and pests, full of unadulterated goodness with vitamins, minerals, enzymes and other nutrients intact and in balance. So why should preparing for a healthy baby not follow the same successful principles?

If both parents are in optimum health when they conceive, their offspring are likely to be as well. If, conversely, the parents have neglected or even abused their own health, are overloaded with toxins from eating chemical junk food, have vitamin, mineral and essential fatty acid deficiencies and bodily systems or organs not functioning correctly, then these problems are likely to severely disadvantage the baby's health and probably cause major problems in either infancy or as the child matures.

Pre-conceptual care is the term given to the concept of bringing both parents as close to optimum health as possible before they try to conceive, and creating the healthiest physical environment possible for a growing embryo and foetus. Why disadvantage your baby's health when you have the opportunity to go organic, detox and clean up your diet, restore your health, green your house and lifestyle, and lay the

---

Foresight survey 1997-1999

Out of 1076 couples participating in the survey, 1,061 had previous fertility or miscarriage problems. 729 couples conceived (67.75%). The couple gave birth to 779 babies, and among the couples there were 67 who were pregnant when the survey ended. If these are added in, it makes a success rate of 78.4%.

Among the 1076 participating couples there were 393 who had previously suffered miscarriage. From this group, there were 28 miscarriages. From those who hand not miscarried previously, there were no miscarriages.

From the 779 conceptions, the expected rate of miscarriage would be 218.5 (25%), but the actual rate for those on the programme was 3.2%. Among women who had previously miscarried, the expected rate would have been 131 (33%), but the actual rate was 7.1%.

The rate for malformations in pregnancies calculated by the NHS is 1 in 17 (6%); for those on the programme, there were only 4 malformations in 846 pregnancies (0.47%).

According to the research, from the basic Foresight programme it is possible to:

• More than double the success rate of IVF (22.6% to 47.1%)

• Raise the success rate of infertile couples overall from 22.6% to 72.4%

• Drop the miscarriage rate from 33% to 7.1% with those who had previously miscarried

• Drop the miscarriage rate to zero for those who have not previously miscarried

• Drop the stillbirth rate from 1.37% to 0.47%

• Drop the malformation rate from 6% to 0.47%

• Drop the low birth weight rate from 9% to 4.6%

foundations of health before you conceive, so that your baby will be resistant to disease and illness? After all, your baby will literally be formed from you and your partner's bodily tissues and blood, and from what you both consumed and were exposed to before and during pre-conception months, and during your pregnancy. And pregnancy, birth, coping with new motherhood, and breastfeeding all make serious physical demands on your body; they are likely to be easier, more positive and rewarding if you're in top shape.

Pre-conceptual care is not a new concept, but unfortunately in this country it is still relatively unknown and is not widely practised. Its benefits, however, are documented in recent UK research which shows that by ridding our systems of harmful toxins, healing infections, reducing or eliminating allergies and vitamin, mineral and essential fatty acid deficiencies, eating an organic wholefood diet, filtering drinking water and taking the right supplements, it is possible to eliminate most infertility and the need for expensive IVF treatment, and hugely lessen the incidence of foetal deaths and deformities.

This research is based on the renowned Foresight pre-conception programme, which has helped tens of thousands of couples around the world to conceive healthy babies, including those who have not succeeded with IVF, and others whose previous children have been malformed, miscarried or stillborn. The charity's advice is backed up by hundreds of international scientific studies.

Foresight started advocating pre-conceptual care in the 1970s, recommending an organic wholefood diet as an essential part of its programme. The charity's practitioners, such as Antony Haynes of The Nutrition Clinic in London, say that Foresight babies, according to their mothers, are healthier and bigger than other babies around them, learn to read and walk before other children, and seem to be many months ahead in their development. The babies are also reported by their mothers to have less allergies, intolerances and colic, and have better sleeping patterns. (This is also the common experience of parents interviewed for this book, many of whose quotes are featured in Chapters One to Three.)

Foresight provides prospective parents with research evidence, information leaflets, books, booklets, a poster and a video on how to plan for a healthy baby, and offers a pre-conception health programme, including mineral analysis of hair samples to determine deficiencies, and a supplementation programme to correct them. Couples are given the name of a Foresight practitioner and a nutritionist in their area, and further tests may be carried out to detect allergies, infections and any other condition which may affect proper development of the baby.

An organic wholefood diet for both parents forms an important part of the Foresight programme because of the food's apparently increased content of minerals, enzymes and vitamins (Foresight quotes US research showing organic food to be higher in these than non-organic—sometimes by up to 100 times), and lack of artificial toxic chemicals and fertilizers.

As part of its basic plan for six months before conception, Foresight stresses that both parents should "eat a good wholefood diet, free from dangerous additives, and organically produced". It highlights the dangers of environmental toxins to which we are commonly exposed, claiming that there are at least 200 man-made chemicals for which there is published evidence of reproductive hazards, including pesticides, carbon monoxide, formaldehyde, heavy metals and solvents.

Its publications include a comprehensive guide (*Planning for a Healthy Baby* by Belinda Barnes and Suzanne Gail Bradley), which outlines the Foresight programme, and explains why organic food production is better than conventional for both human health and the environment. Barnes and Bradley list several studies showing that foods produced by ecological methods contain more nutrients than those grown by agrochemical methods, and discuss classic health research findings carried out as far back as the 1930s by Drs Weston Price and Francis Pottenger of the US, and Sir Robert McCarrison of the UK, demonstrating the need for unadulterated, nutritious diets in order to produce healthy offspring.

Foresight founder Nim Barnes says that interest is increasing in Foresight's pre-conception care all the time, from parents, alternative health practitioners, midwives, health visitors, nutritionists, doctors and

"Malnutrition in the womb is reflected in low birth weight, and often associated with increased infant mortality and illness in younger life. Professor David Barker of Southampton University (author of *The Foetal Origins of Adult Disease*) found that babies with low birth weight were more likely to suffer high blood pressure, cardiovascular disease and diabetes, and that they had a higher incidence of early death. He suggested that malnourishment in early life, both pre-natal and post-natal, would have profound effects on adult life. We should not be focusing our public health on interventions in adult life because it is too late; we should be focusing on children, not just for their health while young, but long-term. Our aims should be to optimize maternal nutrition because that is a way of ensuring the foetus is well nourished. We should strive to optimize foetal nutrition indirectly by improving mothers' nutrition, to optimize infant nutrition by making sure babies are breastfed and receiving the best weaning foods, and to optimize child nutrition by good food at home and at school."—Professor Lawrence Weaver, Department of Child Health, University of Glasgow. speaking at the London Food, Children & Health Conference, 1998.

gynaecologists, and that the increased availability of organic food—especially in supermarkets—has made it easier for parents following the programme.

"The government should be encouraged to subsidize organic food production," she says, "as they are currently subsidizing people to produce inferior food."

Foresight, which also has branches in the US and Australia, carried out a survey of the progress of 1,076 mothers who followed the programme between 1997-1999. The results of this research showed that out of the 1,076 couples (1,061 of whom had had previous fertility or miscarriage problems), 729 couples conceived (67.75%), and gave birth to 779 babies. There were another 67 mothers who were pregnant at the end of the survey; if these are added in, it makes a success rate of 78.4% (*see box on page 40 for further results from the survey*).

Some couples carry out the Foresight programme and also use high-tech procedures such as IVF. However, another pre-conceptual care proponent, Patrick Holford, founder of the Institute for Optimum Nutrition, believes that IVF should be very much a last resort. "Our experience is that when people improve their nutrition and lifestyle to maximize their chances of getting pregnant, often IVF becomes unnecessary. The choice of IVF doesn't always mean that the underlying causes have been properly examined first."

And one obvious problem with using donated eggs or sperm—aside from the costs and drugs involved with IVF—is that the health of the donors is unknown, and they could have been heavily exposed to artificial chemicals, tobacco, alcohol, cannabis and other drugs or toxins.

In his best known book, *The Optimum Nutrition Bible*, Patrick Holford claims that the months before conception and during pregnancy are the most critical period of life, that scientists are increasingly discovering the profound effects of the mother's health and nutrition on the infant during this time, and that adult disease patterns can be traced to infancy nutrition. Optimum nutrition incorporating a mainly raw organic diet, he says, increases fertility, health during pregnancy and the chance of producing a healthy baby with resilience to disease.

A recent book by Dr Marilyn Glenville, *Natural Solutions to Infertility: how to increase your chances of conceiving and preventing a miscarriage*, published by Piatkus Books, describes a four-month pre-conception plan based on her work with Foresight and the Hale Clinic.

"In our desperation for a baby, after years of fertility treatment, we thought going organic might help our next attempt at assisted conception. Our first child was born later that year!"—Fran Davies, Suffolk.

Lose the toxic fat

Also promoting the benefits of an organic diet before conception, pregnancy and breastfeeding (along with the need to rid the body of contaminated fat containing toxins accumulated over the years), is Dr Michel Odent, French author of *Primal Health*, promoter of natural childbirth and of the use of birthing pools, and founder of the Primal Health Research Centre, a London-based charity. Dr Odent set up the Centre in the early 1990s to raise awareness of the very early causes (i.e. in the womb and during the first year of life—what he calls the primal period) of disease and health degeneration which affect us in later life, and to offer practical solutions.

In September 1999, the centre started Dr Odent's new Accordion Method of renewing body fat in women planning to conceive, in order to significantly reduce or stop the transfer of pollutants stored in body fat from mother to baby via the placenta and breastmilk. The Accordion Method involves a series of short semi-fasting sessions so that a fast weight loss is immediately followed by a fast weight recovery. Steam baths, physical exercises, a special Swiss formulated juice and lectures to occupy the mind and fully explain pre-conception care, healthy pregnancy and breastfeeding, are all part of the three-day programme which can then be continued for up to a week at home. The programme can be repeated twice a year, or every few months if shorter fasts are undertaken. The juice used in the fast is a Swiss-made formula of special palm tree and maple syrups combined with fresh organic lemon, pure water and cayenne pepper.

"When I had children I started thinking about what we ate—especially with the BSE crisis. I had previously been involved in poultry production and knew they fed offal to the chickens. I have been concerned that with school dinners, the cheapest, low grade meat is bought for children, which is most likely to be contaminated by any diseases. Now being organic is the only route in my opinion. It's expensive, which is unfortunate, and one of the difficulties is that you need to shop in different places to get all the food you need. I make sure that all the basic staple items—which aren't expensive— are organic, such as milk, pasta, butter, rice and eggs. We mainly avoid imported, and buy in season. We forgo new clothes, magazines, and some outings in order to eat organic, and I believe it costs us £60-£70 a month extra. Buy as much organic as you can afford, and locally produced if possible, and learn about what's in season throughout the year. You'll save on doctors and medicines."—Helen Pitel, North London.

Dr Odent believes that it's the only logical strategy—unless you run regular marathons—to reduce the body's accumulated toxic load before conceiving and thus avoid passing on toxins to a baby. He also advocates an organic diet, avoiding health-damaging transfatty acids (abundant in cooked processed oils, hydrogenated fats and oils, margarines, chips and many other fast foods, and which can be passed on to the baby), limiting fatty dairy products, choosing lean meat over fat meat, avoiding the skin on poultry, and avoiding freshwater fish and fish high up the food chain, as these are believed to be more polluted. Dr Odent also warns mothers against losing weight too fast during breastfeeding, as this can release toxins into the breastmilk—creating what he describes as the 'slimming nursing mother syndrome', after hearing breastfeeding mothers claim their baby gets sick whenever they lose weight. (Losing weight deliberately during pregnancy and breastfeeding is not recommended and could be dangerous for your baby.)

Dr. Odent has teamed up with Dr Vyvyan Howard, a well-known expert in foetal and infant toxico-pathology, and hopes to get lottery funding for a proper trial of the Accordion programme in 2003.

How soon is soon enough?
The purist's view

While organisations such as Foresight and many leading practitioners recommend pre-conceptual care from between four and six months to one year before conception, it could feasibly take several years to fully detox, to clear a mother's body of most man-made chemical pollutants and ensure she does not pass these on via the placenta or breastmilk in harmful quantities (and fully restore all bodily systems and nutrient levels). Most of these chemicals are not biodegradable, and are stored long-term in our body fat: can we really lose most of them in short fasting sessions? Also, let's not forget the father's important role in providing healthy sperm. It also takes some time to fully rejuvenate a body previously overloaded by a refined, highly processed diet, when moving on to fresh organic wholefoods. Health expert Leslie Kenton believes that once the body's balance has been disturbed by a deficient diet, restoring it is a slow process taking many months, even years.

So from a purist's point of view, prospective parents should allow several years from the day they start to fully detox and go organic, to the time they try to conceive. After all, to satisfy the UK organic standards which are applied to all certified organic foods in this country, a farmer spends at least two to three years converting land, crops and animals from chemically based agriculture, before the produce can be considered for certification. So for a truly organic baby, and to have real peace of mind, going mostly or totally organic with the emphasis on mainly raw wholefoods at least two to three years before conception would be ideal (or possibly even longer, as many chemical pollutants are non-biodegradable, are accumulated over our lifetime, and stay in body fat for many years).

As part of this process a lengthy, professionally guided detoxification programme would be needed, including extensive fasts to lose contaminated body fat well before conception. Other measures can include colonic irrigation, skin brushing, deep massage, aerobic exercise, yoga or Pilates, and increased pure water intake to aid fat loss and allow the liver, lymphatic system and bowels to be fully flushed of all toxins. Replacement of mercury amalgam fillings could be considered (the mercury is believed to be able to leach into the bloodstream, causing health problems for both mother and child, and fillings are therefore officially not recommended to be carried out during pregnancy). Once cleansing was complete, adequate healing time (a period of many months, even a year or longer) would be required to restore body fat, correct deficiencies, fully replenish the body with nutrients and create a sufficient health bank for baby, from a wide variety of mainly raw organic foods with supplementation as necessary, before trying to conceive.

However, we don't live in an ideal world, so going organic whenever and however you can—even long after your baby is born—will still significantly benefit him, and you, now and in the years ahead.

What you can do if you're planning a baby

• **Go organic NOW!** Time is on your side, so start building up your health—and that of your partner—by eating a mainly organic wholefood diet, ideally with as much raw, steamed or lightly cooked food and pure filtered water, as possible. Take your time to find a good, reliable and regular source such as an organic box scheme or organic home delivery company. The sooner you start to eat food containing more nutrients and avoid exposure to artificial chemical residues the better for you and your baby-to-be. Explain your new diet and lifestyle to friends and family, encourage them to have organic food when they visit you and when you visit them, and try to go to organic eateries when you venture out for a restaurant meal. Follow the relevant Organic Food, Health & Nutrition Tips in Chapter Three. With organic foods, look for the certifying symbol, such as those of the UK Soil Association, Biodynamic Agricultural Association (Demeter), Organic Farmers & Growers and the Organic Food Federation.

• **Follow the steps on how to convert** if you already have children, or the suggestions for how to be an organic parent in Chapter Three.

• **Send an A5 SAE with a 33p stamp to Foresight,** 28, The Paddock, Godalming, Surrey GU7 1XD, to receive general information on this registered charity's excellent pre-conception recommendations. Foresight, which offers annual membership for £21.15, will recommend that you have a hair mineral analysis and possibly other tests such as urine, stools, blood and saliva to pinpoint any vitamin and mineral deficiencies, allergies or infections. These can then be corrected before conception. As part of its programme, Foresight recommends that you give up tea, coffee, alcohol, cigarettes and drugs, and immediately eliminate or reduce exposure to any environmental toxins, properly filter your water, and eat a mainly organic wholefood diet. Natural birth control methods are also recommended. Foresight vitamins, minerals and other useful pre-natal supplements are available in health stores, and all profits go towards Foresight's important work. An excellent book detailing the programme is available from the charity (*Planning for a Healthy Baby* by Belinda Barnes and Suzanne Gail Bradley), and the *Foresight Wholefood Cookbook* shows how to prepare food without losing vital nutrients.

• **Alternatively (or in addition), you can try the Accordion Method** advocated by Dr Michel Odent of the Primal Health Research Centre in London to lose toxic body fat before conception, and follow his other dietary advice; or follow a strict detox diet and rejuvenative health programme well before conception, such as the seven-week general programme devised by London naturopath Kitty Campion. She uses Vega

"The more I read about additives, pesticides and antibiotics, etc, the more I wanted to avoid them, and we decided to go organic when weaning our first baby. I like to think that by using organic products where possible I am helping my own and my family's future health and that of the environment. I feel better somehow, knowing that I am making a contribution, however small, towards a more healthy world. I definitely don't use disposable nappies on my daughter as I was horrified to learn of their chemical content and their impact on the environment, and instead use a cotton nappy service. The cost of organic food has been an issue for us and I have felt unable to justify buying certain organic products when the non-organic version is so much cheaper. However, I have prioritized by always buying organic meat, eggs and most vegetables. My advice to other parents thinking about going organic is to decide what you can afford and then prioritize—even if you only use some organic products, it has to be better than using none at all."—Karen MacMillan, Berkshire.

analysis, complementary therapies, colonic irrigation, raw food, herbs and superfood supplements. Seek professional advice and guidance throughout such programmes (see Leading Practitioners in Part Two). You could consult other good pre-conceptual books for their recommendations, such as *Healthy Parents, Healthy Baby* by Maryon Stewart and *Holistic Herbal for Mother & Baby* by Kitty Campion (see Publications in Part Two). You could try the practical steps to restored health outlined in many of Leslie Kenton's excellent books including *Ten Steps to Energy* and the *Raw Energy* series, or go to a nutritional workshop or course such as those run by Daphne Lambert at Penrhos Court, or at the Institute for Optimum Nutrition (see Part Two). See also *The Organic Baby & Toddler Cookbook*.

• **Find a good qualified naturopath or nutritionist** to ensure that your diet is well balanced (filled with the correct proportions of nutrients you need for yourself and your baby-to-be such as zinc, vitamin C, essential fatty acids etc, which are vital for conception and for your baby's proper development) and help to build a sufficient organic health bank for both you and the baby to draw upon during pregnancy, birth, breastfeeding and beyond. According to Dr Sarah Brewer in her book *Super Baby*, when you are pregnant, over half the nutrients you eat are used by your growing baby. Consider starting to take high quality pre-natal vitamins and minerals—see the Products Guide and seek professional nutritional advice. Ideally, continue to seek this advice and support throughout pre-conception, pregnancy, breastfeeding and weaning. Ask Foresight for your nearest clinician and nutritionist, approach the Institute for Optimum Nutrition, the Naturopathic Helpline, or practitioners listed in Part Two.

• **Consult the Products Guide in Part Two for organic 'superfood' supplements** which can help cleanse and enrich the body ready for conception (check with your natural therapist, as some products may not be suitable to be continued during pregnancy). They include Progreens,

"Organic foods are tastier and we feel confident that we're eating healthy food produced with respect for animals and the environment. The cost of organic food has been an issue for us, however prices seem to be falling and now some organic foods are only slightly more expensive than conventional brands. Initially it was the BSE crisis that made us think carefully about our food and we have concerns about modern food production methods including factory farming, the use of pesticides and antibiotics, and genetically engineered foods. If you also feel passionately about the issues of food production please write to or phone companies with your concerns. As a busy mother I don't always have time to write, but often get results from a quick phone call."—Mrs Lewis, West Glamorgan.

Foresight Vitamin C and Garlic, and separate Foresight Vitamin and Mineral supplements (a programme of nutrients will be tailor-made for you by Foresight after your hair analysis), Seagreens organic seaweed capsules (seaweed can help prevent or reduce fluid retention, and its nutrients are easily absorbed by the body), Pure Synergy and Living Food Energy. Vegetable and seed oils such as Udo's Choice and Essential Balance are ideal sources of essential fatty acids.

• **Get fit**—your body is going to go through extreme physical stress with pregnancy, birth and early parenthood, and being fit, healthy and well oxygenated is essential to keep you mobile, flexible (important when you're nearly at term) and vital, and keep you sane! And pregnancy is not the time to take up vigorous exercise. Walking, swimming, yoga, Pilates, cycling and rebounding (using a small, approved trampoline) at your own pace, are all ideal.

• **Practise breathing deeply and take time out to relax**. A warm bath with a few drops of organic essential oils works wonders (but check which ones—some are unsuitable for use during pregnancy). Use creative visualisation techniques: imagine you're on a luxury holiday on a tropical island and have a half hour 'holiday' without leaving home, enjoying all the imaginary sights, sounds, smells and feel of the water and beach sand.

• **Use complementary therapies** to heal your body, prepare it and help it through conception, pregnancy and birth, such as homeopathy,

---

"It's so important to have an organic diet while pregnant. My daughter Saba and eats 90% organic, and her immune system is like the Great Wall of China—nothing gets through. She caught chicken-pox and we nearly didn't even notice—she had a couple of spots. She's never been vaccinated—it makes no sense to do so, especially as I avoided all drugs during her birth—and there was no way we would have her done, and I've travelled with her through Asia and South America without illness. She eats healthy organic food such as brown rice and vegetables, which is the cheapest way to eat organic, and we don't buy many ready made foods. If Saba does have sweets on the odd occasion they're organic. She's always been so alert and full of energy, and has always been off the scales, is very tall and robust. She doesn't get colds—meanwhile her friends are an endless stream of snot during the winter. And she really understands the organic message—if I tell her a food's not organic, she doesn't want it. And we don't have supplements, I believe that if you eat a good organic diet you shouldn't need them as the vitamins and minerals are in the food—an organic apple apparently has many times the vitamin C of a non-organic one."—Tania Smith, Brighton.

osteopathy, massage, Reiki and aromatherapy. The Hale Clinic in central London offers all of these services, as does the Birth Unit at St John and Elizabeth Hospital in St John's Wood, or find practitioners near you.

• **Get lots of fresh air** away from polluted city streets. If you must be around city traffic, consider wearing a mask filter (replace the filter frequently—you can get them in most bike shops).

• **Avoid all harmful man-made chemicals as much as possible,** for example glues, typing correction fluids, marker pens, cigarette smoke and other environmental toxins and hazards at work. If you or your partner come into contact with these as a necessary part of your job or hobby (for instance if you are a spray painter, photographic laboratory technician or photographer with home darkroom, work in a nuclear power plant, chemicals factory or similar), consider changing your job or taking extended leave, or changing your hobby. Reduce your use of VDUs, mobile phones, etc.

• **Green your house and garden**: reduce your exposure to harmful chemicals and contaminants in the home and garden by changing to environmentally friendly cleaning agents and paints; and avoiding aerosols such as air fresheners and fly sprays, varnishes and polishers, artificial bodycare toiletries (including deodorants containing aluminium chlorohydrate and fluoridated toothpaste with artificial sweeteners), pet, plant, pest, timber and lawn chemicals and fertilizers, etc. Get a plumbed-in water filter (they can be rented) and, if you can, investigate using a shower filter as well, to avoid chlorine and other chemical absorption during showering (see Part Two for products and stockists). Consider environmental hazards surrounding your home—for instance factories, non-organic farms, overhead power lines and landfill sites.

"There is so much confusion and counter advice heaped on parents now that it can be quite paralysing. One safe and sure thing that parents can grasp on to is that if they buy organic food it's the best shortcut to getting the best food possible for their baby. If I had another baby now I'd be delighted at the prospect of being able to feed that baby completely organically because I believe it would give the child the best start in life. Buying organic is the best way of sourcing food which you know has been produced as naturally as possible, which hasn't caused animals to suffer, which hasn't trashed the environment, and which is likely to have good levels of nutrients because it hasn't been persuaded to grow artificially with endless fertilizers."—Joanna Blythman, author of *How to Avoid GM Foods* and *The Food Our Children Eat.*

• **Wear organic and/or untreated clothing** (try to avoid clothes which need dry cleaning, as fluids used in this process have also been found to contaminate breastmilk), and try organic and untreated bedding. If you buy a conventional mattress and bed base, let them offgas and air in a separate room for several days. See Part Two for products and stockists.

• **Keep away from farms, golf courses, school fields and parks during or immediately after spraying,** and check with your local council to find out if any spraying is to be carried out in your neighbourhood. Ask neighbours not to use chemical treatments in their gardens, and help them with alternatives by putting them in touch with the Henry Doubleday Research Association (HDRA), the organic gardening authority, or by lending them a leading organic gardening book.

• **Have a dental checkup,** and if you need to have new fillings or work done on existing fillings, have these done many months prior to conception to avoid mercury contamination (the British Society for Mercury-free Dentistry holds a list of their members). Purists would advise having any mercury amalgam fillings removed and replaced well before conception.

• **Avoid having perms** or any other hair treatments using peroxide, ammonium or other strong chemicals, and don't use hair-sprays or gels.

• **Use non-toxic materials** and paints if you're decorating the nursery ready for your baby. For example avoid using or cutting MDF (medium-density fibreboard, used as a wood substitute and containing many solvents), oil-based paints, etc.

• **Avoid taking antibiotics,** prescription drugs and other pharmaceuticals, and avoid x-rays.

• **Start thinking about the issue of vaccination** (see section in Chapter Three) and whether you want your baby to have vitamin K at birth. Read up, talk to practitioners and other mothers to find out all you need to know in order to make an informed decision that's right for you and your family. Don't be pushed or rushed into a decision.

Where to start if you're already pregnant

• **Congratulations!** You still have the chance to give your unborn baby a head start in life. It is important, however, that you act to go organic NOW, and eat and drink as much organically from here on in as you can. Get the support that you need from relevant publications, natural therapists and other parents and parents-to-be. Set up a weekly home delivery of organic fruit and vegetables today, or get them and organic groceries from a similarly reliable source, and buy or rent a water filter (buy bottled water in the meantime). Follow the relevant Organic Food, Health & Nutrition Tips in Chapter Three, and the suggestions for a fulfilling organic parenthood. With organic foods, look for the certifying symbol, such as those of the UK Soil Association, Biodynamic Agricultural Association (Demeter), Organic Farmers & Growers, and the Organic Food Federation.

• **Follow all but the first four points in Pre-conception** (from 'Find a good qualified naturopath or nutritionist' on page 47, to the end of the section).

• **Start buying for your baby**—many organic mail order products can take several weeks to arrive, and if you're ordering for a newborn you don't want to be caught short if your baby arrives a few weeks early. Make good use of the Products Guide to request suppliers' catalogues, or visit some of the stores now selling organic goods. It's also a good time to catch up on reading, and there are many useful publications featured in the guide. There are also interesting and useful organisations you may wish to join. Don't be afraid to ask family and friends to buy organic items for your baby—especially if you're on a tight budget and can't afford to buy all the ones you want. Put together a wish list and pass it around.

Breastfeeding
Why organic breastmilk is best

Breastmilk is the food that nature designed for babies. Nothing, not even the most scientifically formulated infant formula or follow-on milk, can equal it for giving a newborn all the essential nutrients it needs for optimum development until at least four to six months of age. Breastfed babies have been shown in studies to have higher IQs than bottle-fed babies, because of the essential fatty acids and other key nutrients in breastmilk. And international scientists have hailed breastmilk as a possible treatment for cancer after discovering it kills cancer cells.

Ideally, babies should be breastfed on demand until at least six months to one year of age, and longer if possible. It is the most natural and healthy

"Formula milks based on milk from another species, such as the cow, are deficient in many good nutrients for infants. We are recognizing more and more clearly that cow's milk is for calves and human's milk is for babies. For the last hundred years we have been conducting a controlled experiment in bringing up children on the modified milk of another mammal, and we're just beginning to recognize the long-term effects."—Professor Lawrence Weaver, Dept of Child Health, University of Glasgow.

way to feed your baby. There are many concerns about introducing infants to another animal's milk proteins, such as those in cow's milk, at such a young age, including the increased likelihood of allergies and asthma. It's also convenient, as there's no need to buy expensive formula, or to heat, sterilize and wash bottles, especially in the middle of the night.

It is incredibly sad that in this country very few mothers breastfeed their babies beyond one month, and many unfortunately decide they don't even want to try, and instead bottle-feed from birth. It has been found that just over half of all UK mothers are breastfeeding their babies two weeks after the birth, and only just over a quarter are still breastfeeding at four months. Of course, some mothers do experience problems and aren't able to breastfeed, but for those who have a choice, breast is certainly best for your baby's health and development. Babies born prematurely and in special care can be fed with expressed breastmilk, which greatly increases their chances of survival—they can progress to normal breastfeeding when strong enough. There are now several organisations such as the National Childbirth Trust and various parent networks offering advice, support and equipment such as electric breast pumps for women having difficulties breastfeeding, or wanting to increase their milk flow. They are listed at the back of the book, so make good use of them and try not to give up too easily if you do encounter any hurdles—both you and your baby will be glad you persevered in the long run.

If the mother is eating a well balanced, mainly fresh and raw organic diet, is supplementing as necessary and is not stressed, and certainly if she has been on such a diet for some time (i.e. during pregnancy and before conception), then the organic breastmilk she produces should be of premium quality, and any pollutants should be significantly reduced. The longer the mother has been eating organically and reducing her exposure to environmental toxins the better, as studies on human

"According to a study recently published in *The Lancet* which compiled data from 150,000 women in 30 countries around the world, breastfeeding can reduce the risk of breast cancer. The report also showed that each child's birth increases a woman's level of protection against the disease. This explains why breast cancer is not so prevalent in developing countries where mothers have an average of six children, each of whom is breastfed for up to two years."

"Children who are breastfed are less susceptible to many illnesses such as chest or middle ear infections, asthma, eczema, diabetes and gastroenteritis. They are less likely to be obese and studies have shown that they score higher in IQ tests."

—from *Green Parent* (www.thegreenparent.co.uk)

breastmilk have highlighted residues of pollutants in both the milk and body fat. You can renew your body fat before conception, pregnancy and breastfeeding to further reduce the pollutants passed on to your baby by undertaking a series of fasts (see 'Lose the toxic body fat'); ironically, the longer you breastfeed, the less residues are left in your body. As toxins are released during fast weight loss, it is not advisable to slim during breastfeeding; it also reduces essential nutrients and fats required by the baby for its growth and development.

All in all, there are many benefits of breastfeeding for you and your baby, and there are even books which are entirely devoted to the subject, such as *Breast is Best* by Dr Penny Stanway. Breastmilk gives your baby increased resistance to illness and disease, and helps to ward off infections by passing on antibodies. It also reduces the risk of developing allergies. It is especially important for your baby to be breastfed for as long as possible if he is not being vaccinated, as it helps to build the immune system and protect against disease, but many would also say that it's as important to breastfeed a vaccinated baby, in order to help him cope with and recover from the vaccinations.

It is also extremely rewarding to be able to feed your baby entirely from your own body, emotionally and physically bonding for you both, and extremely convenient—there's no time spent preparing feeds or heating them in the middle of the night, and you can feed your child anywhere, any time he needs it. You can also express milk and freeze it

"About a year before I became pregnant I became concerned about toxins and pollution in food and increasingly interested in organic food, and have eaten it ever since. My baby is now eight months and has been organic since weaning. Compared to other babies he's big for his age and has less colds, and recovers quickly from them. He's very strong and more physically developed than other children and has always been very alert. Some people think I'm weird by only feeding him organic, and ask why I won't give him cheesy wotsits—but I don't want to give him that sort of stuff, I want him to have proper food and give him the best start. When he's a teenager he can eat what he wants, but while he's little I'll give him a good start with organic foods, and filtered water. I certainly have peace of mind knowing that I'm minimizing the amount of toxins and chemicals entering our bodies, and as organic food preparations tend to have less fat, be lower in cholesterol, no artificial E numbers or additives, our diet is much better and we feel more energetic. I'm prepared to pay a bit more for these healthier foods and it's becoming easier to shop organic due to increasing stock and wider ranges in shops. It's especially important for mothers to go organic for breastfeeding, as babies are taking in a lot of milk and nutrients from the breastmilk."—Josephine Williams, South London.

in special bags or containers (made from an inert plastic or glass) for use in weaning foods, or to ensure a supply of favourite food if you're away from each other for several hours.

Many large retail stores such as Mothercare, the Early Learning Centre, Boots, department stores and supermarkets provide special breastfeeding areas if you need them—so make use of them when you are out and about. Don't allow yourself—or any other mother—to be evicted from a restaurant or public place for breastfeeding—stand up for mothers' and babies' rights! No one should expect a mother to have to feed her baby—or the baby to have to eat his dinner—in the ladies' toilets.

Breastfeeding can be continued when working—especially easily if you work from home, or if you are fortunate enough to have a nanny or carer who can bring your baby to you for feeds if you work nearby. Alternatively, if you are able to get your baby to take a bottle, you can express your breastmilk and freeze or refrigerate it ready for someone else to feed to him. Dawn Thomas, an Agenda 21 Co-ordinator in Kent, returned to work when her baby was seven months old and expressed during work and at night to keep her son going. She says it's an effort to express the 15 ounces needed and finds she has to be determined, but that it can be done. She plans to breastfeed her son until he is at least a year old while working full-time. "Talk to your employer and explain why you want to do it, and make the necessary arrangements—such as having a private area to express your milk, and somewhere to clean your equipment and refrigerate or freeze your milk. It's tiring, and you have to take breaks, watch what you eat and have plenty of fluid, but it's been worth it for the health of my son."

Dawn advises gearing your body up to expressing well beforehand if you are planning to to do so when you return to work, and building up a stock so that the first day or two isn't critical, and you don't get stressed.

"A variety of studies have shown the beneficial effects of breastmilk on a baby's emotional, physical and intellectual development. By the age of 3 months, the IQ of babies who are breastfed is 3 points higher than those fed on formula, and those who are breastfed until 6 months have an IQ that is 6 points higher than those receiving formula. Some studies suggest that babies who are breastfed for at least 6 months enjoy a 10-point IQ advantage over non-breastfed children when assessed in later childhood. Even breastfeeding for just 3 weeks can improve your child's long-term IQ by an average of 4 points, as it supplies important building blocks for the brain at the time when it needs them most. Research (also) shows that breastfeeding improves hand-eye co-ordination, visual development, language and social skills."—Dr Sarah Brewer, *Super Baby*

Chapter 3

Organic parenthood
A rough guide

NEW PARENTHOOD is a steep learning curve, and a time of serious readjustment, whether you're prepared for it or not. Just as you're coming to terms with the biggest upheaval in your life, you're also making daily decisions about all kinds of new products, services and ways of doing things which are unfamiliar. And believe me, there'll be no shortage of advice from people (midwives, health visitors, GPs, mothers and in-laws), and from books and magazines—all trying to tell you what's best for your baby, what to do and what to buy.

It's the ideal time to decide that whatever parenthood throws your way, you'll handle it in a way that's as environmentally friendly and healthy as possible, by choosing organic first: organic food, organic baby-gros, cot blankets, nappies and even soap and shampoo if the family budget allows.

There probably won't be a lot of spare time or energy available in the first few months of your new role, so prepare well in advance so that you can make an informed choice about baby products, get the ones you want when you need them, and not have to settle for undesirable substitutes at the last minute. Buying organic may not always be the convenient choice—most of us don't have the good fortune to live close to a good organic farm or wholefood shop, let alone an organic supermarket or organic baby shop. And even if you do, there isn't a one-stop organic store in the country which has all the food and products you'll require, so you'll need to shop around. Most are readily available by mail order however, which can be extremely convenient.

There is extra effort involved in being an organic parent, but anything that requires effort is usually worth it—your baby for instance! You'll be able to sleep contentedly at night (baby permitting), knowing that you are doing the best you could possibly do for your child's health and wellbeing, as well as for the rest of your family, the environment and wildlife (and other families, especially if you buy fair-traded organic goods).

Organic parenting tips

Where products, services, organisations and publications are suggested, refer to the relevant sections in the Products Guide. A number of particularly useful books are listed under Food and Health Publications.

• **Buy and eat locally produced, fresh organic food** for your family wherever possible—to ensure that your baby's food is grown without harmful chemicals and GMOs and has maximum nutrients, that 'food miles' are reduced and that your bodies are 'in synch' with the seasons. Choose fresh rather than processed and packaged. Ideally, use a local organic farm box scheme or farm shop to ensure maximum freshness of food, to support your local economy and reduce food miles. Farmers' markets are also an ideal choice, as the farmer is bringing his freshly harvested food straight to the consumer. If these options are not practical, choose a regional or nationwide organic home delivery service. All of these use minimal, recyclable packaging, and supply food as freshly harvested as possible. In contrast, much supermarket produce is wrapped (sometimes double-wrapped) in plastic, which is undesirable from both an environmental and health point of view (goods are often transported around the country purely for the purpose of packaging).

• **Make your own fresh and mainly raw, steamed or lightly cooked organic babyfood** (see Organic Food, Health & Nutrition Tips) to get maximum nutrients into their tiny bodies, and at the same time eliminate packaging waste and save money. Keep a supply of frozen organic vegetables (which can have more nutrients in them than so-called fresh, because they are frozen immediately after harvesting), and frozen ice cubes of leftover food in the freezer to use as a back-up. Use top quality fresh, jar or packet organic ready-made babyfood as another back-up, or for when travelling. Get your baby on to your family meals as soon as possible, perhaps starting with a small blend of your dinner from age six to eight months. Follow the suggestions in Organic Food, Health & Nutrition Tips.

• **Buy as many of the organic versions of products** that you need for your baby as you can reasonably afford (bear in mind that some suppliers will let you pay over several months). With food, ensure that at least all the major staples—the groceries you use the most of, such as milk and bread—are organic. Make sure any food intended solely for your baby is organic, especially ready-made babyfood, as multiple pesticide residues have been found in conventional babyfood meals. If you can only afford a few organic clothing items, buy good quality ones that your baby won't grow out of so quickly, and especially the ones in which he'll spend the most time and are closest to his skin (when organic baby-gros get too short, you can always cut the feet out of them). Buy organic cotton sheets, cot blankets and other bedding which will last at least for the first year—with luck, longer—to make them more economical. Better still, buy the organic material and make your own! (A rewarding project during the latter stages of pregnancy, when you're fed up with waiting.)

• **Buy reusable organic cotton nappies** for your baby, and wash them using environmentally friendly or organic cleaning agents. Use a nappy service if the washing overwhelms you (find out what cleaning agents they use). If you must use disposables at any time, such as for long-distance travelling, use non-gel, non-plastic ones which can be composted, such as the Tushies brand. See the section in Part Two on why organic reusable nappies are better for both baby and the environment, and *The Humanure Handbook* in the Publications section. If you find it difficult to move away from using disposables, try a mixed nappy system at first, and rotate your use of reusables and disposables. Slowly build up the number of reusables until you're only using disposables as an emergency back up.

• **If you do buy any non-organic food, make sure it's GM-free**. Avoid all GM foods and those that may contain GM ingredients—soya and maize (referred to as 'corn' in some products) derivatives are among the ones to question, along with all non-organic vegetable oils. Ask the store or manufacturer for a guarantee if you're unsure.

• **Breastfeed your baby for as long as possible on demand**—at least six months and ideally one to two years. Eat a high quality, balanced, mainly raw organic diet with plenty of essential fatty acids so your breastmilk really is best for your baby—a complete food which delivers all necessary nutrients until at least six months of age. Breastmilk quality is crucial to your baby's mental and physical development.

• **Green your house and garden for your baby's arrival** (see The Green Home in the Products Guide). Buy and use organic or environmentally friendly consumer goods, cleaning agents (perhaps consider ecoballs instead of laundry detergents), paints, building materials (avoid or seal MDF board, and have any asbestos professionally removed) and consider using energy-saving devices such as Savaplugs, solar heating, energy-efficient lighting and heating, and perhaps even a household water recycling system. Don't use chemicals and fertilizers in your garden (take an organic gardening course, or read any of the good organic gardening books now available).

• **Avoid fitted carpets**. These can trap toxic dust and dirt, which can cause or aggravate allergies, and there are concerns about the chemicals and materials used in both carpets and underlays. Wooden floors can be swept and mopped. Alternatively, choose natural, untreated floor coverings such as coir or seagrass, available from Fired Earth and The Healthy House (see The Green Home in Part Two). Buy an efficient filtration vacuum cleaner such as the top of the Dyson range, which doesn't re-circulate dust into your house.

• **Avoid bodycare toiletries with artificial chemical ingredients** (i.e. most of those on supermarket or chemists' shelves), including fluoridated toothpastes with sugar or artificial sweeteners, sunscreens, shampoos, soaps and deodorants containing aluminium chlorohydrate. Buy organic and environmentally friendly products from 'green' stockists, and read all labels thoroughly (see Bodycare Toiletries in Part Two).

• **Encourage your child's school to improve the quality of its school dinners** and tuckshop items—get them to include healthy organic foods if possible, and ban all GM ingredients if they haven't done so already. Encourage them not to use chemicals on school lawns and fields (the HDRA has an excellent manual for managing such ground organically), nor to spray or otherwise chemically treat canteens and other areas for cockroaches and other pests.

• **Keep your baby or child covered up or out of the sun**—Australian research has shown that sunscreens do not guard against deadly melanoma skin cancer, and only reduce the risk of other skin cancer by up to 40%.

• **Support the anti-GM lobby in any way you can.** Unfortunately, just buying organic products is not enough, as organic farming as a whole is under direct threat from contamination by GM crops that are being grown in field trials. If large-scale trials and commercial planting go ahead, it could be the end of the road for organic food and farming in this country, as cross-pollination would be inevitable. GM crops are also likely to increase the use of pesticides, which invariably end up in our

> "We chose to go organic about five years ago because 'as ye sow, so shall ye reap'. We feel non-hypocritical, healthy, happy, in control of our lives and have peace of mind as a result. I also feel excited to be part of a positive movement of change in our world. We don't get food poisoning bugs when others do, and have been very well since going organic. We're gradually changing our clothes and bed linen over—most new garments added to the wardrobe are organic and are so comfortable and look lovely and individual. We save up, or wait for the catalogue to have an end of year sale. Buying hemp jeans is a revelation—hemp lasts much longer than denim or cotton, looks like linen, and is soft. We use organic skin and hair products which have no side effects and are luxurious."— Sarah Platt, Brighton.

drinking water and in the atmosphere, and further contribute to the decline of wildlife and biodiversity in the countryside. Organic is GM-free—let's keep it that way!

• **Get fit—create a healthy lifestyle for yourself and your family** with regular gentle exercise and fresh air away from polluted city streets. Gentle aerobic exercise such as walking, cycling (invest in children's seats and helmets), swimming and rebounding using a small, approved trampoline (from the Wholistic Research Company, Fresh Network or Starbound), pumps fresh oxygen around the body, helping to clear out any toxins and rejuvenate you.

• **Read up!** See the excellent organic books and publications listed in Part Two.

• **Join and support organisations** such as the two main organic charities—the Soil Association and the Henry Doubleday Research Association, both of which have local groups around the country—and participate in their organic events alongside like-minded families and individuals.

• **Have a homeopathic family first aid kit** in the house instead of Calpol or other common over-the-counter drugs, and learn how to use it fully: go on a course, or read relevant books such as *Safe, Natural Remedies for Babies and Children* by Amanda Cochrane, *Natural Healing for Women: caring for yourself with herbs, homeopathy and essential oils*, and *Neal's Yard Natural Remedies* (both by Susan Curtis and Romy Fraser), or Jan de Vries' books.

• **Never use organophosphate treatments for head lice** on your children—seek out herbal alternatives such as Biz Niz, or use tea tree oil shampoo. Regular combing is natural prevention, and also spreads natural oils through the hair, keeping it in the best condition.

• **Use health and complementary medicine practitioners** who support and promote organic food and health, such as naturopaths and homeopaths.

• **Tell your health visitor, midwives and GPs** all about the benefits of eating and buying organic for babies and children.

• **Make your own baby wipes** by soaking organic cotton wool pads (such as Bo Weevil's—see Products Guide) in boiled and cooled water in a tupperware container with a few drops of organic lavender oil. If your baby gets nappy rash, try raw organic egg white—it is apparently an almost instant cure.

• **Take your children to a natural dentist** who doesn't use mercury fillings. Many use homeopathics and other natural treatments and products.

- **Teach your children about the importance of a wholefood organic diet** and lifestyle from an early age—make them feel proud of their contribution to a better world. Let them hold and eat whole fruit and vegetables as teeth and co-ordination allow, so they can recognize wholefoods, appreciate their taste, and ask for them. Teach them early on how to peel bananas and oranges, to make eating real food fun. Explain— or better, show them—where food comes from, how it is grown and gets to their plate, so they develop a strong connection with the land. This can easily be done with sprouts or herbs on a windowsill, or potatoes, tomatoes and carrots in buckets on a balcony, your own allotment or veggie patch in your own back garden, and a worm compost bin. Allocate them their own small crop and encourage them to help or fully prepare organic meals. Visit organic farms and let them talk to organic farmers who actually grow their food: the Soil Association has recently launched a network of these. Contact the Henry Doubleday Research Association for information on their courses and to arrange visits to organic gardens, and try some of the excellent organic gardening books now on the

"Increasing concern over food scares was the main prompt for us to go organic, together with having a baby to feed. We started with milk, carrots and potatoes and built up slowly. Now I really think twice about buying a product if I can't find the organic version. We have become much more aware of wider green issues as a result of going organic, such as GMOs, pollution, recycling etc. Despite being a very atopic mother (having hayfever, eczema, asthma and other allergies all my life), both my toddler and baby are clear of any problems. We're very much healthier and enjoy our food much more. Shopping is more interesting and cooking more experimental—there are lots of new vegetables from our local box scheme that I would not have otherwise bought. Cost was a big factor initially, but we realize that having the best quality food is the most important factor in our family's health so now we don't compromise. We buy from a box scheme and supplement in the supermarkets, and waste a lot less as food is so much more precious and tasty. We buy what we need instead of large 'value' packs which go off at the back of the fridge. Also, I now cook a lot more, which saves money, and we now buy almost nothing processed. The knock-on effect from going organic has been huge—we've bought real nappies, eco balls, chemical free household products, a water filter and organic and natural body products. We realize that chemicals and pollutants are everywhere and that we have just woken up from our ignorance! To other parents I'd say don't delay going organic, and don't be put off by the price—not only will it pay dividends in terms of health benefits, but like us, you may well find that the cost balances itself out."—Joanna Wiseman, East London.

market. Buy children's books which follow the organic or environmental ethos. Search out interesting organic sites with them on the internet.

• **Clothe your family in organic (formaldehyde-free and chemical-free) cotton, hemp and linen** wherever possible. Also try using natural, untreated bedding such as organic cotton sheets, mattresses and blankets to eliminate exposure to formaldehyde, solvents, bleaches, strong chemical dyes and other chemicals—we spend a large proportion of our lives in bed, breathing these in. If you can't afford an organic mattress, consider an all-cotton, untreated one.

• **Use second-hand toys, baby books and goods** wherever possible (wash or clean them before use and ensure they are safe), or buy organic and environmentally friendly ones and pass them on when your child outgrows them. Buy wooden toys made from sustainably managed forests wherever possible. Avoid soft PVC toys, which can pass plastic into a chewing baby's mouth.

• **Let any new plastic toys, playnest rings or pools, etc, offgas away from your baby or child** for several days. The same applies to any non-organic or treated mattress or bed base.

• **Don't place your baby near the TV,** or let children sit or play close to the set as there are health concerns about the effects of electromagnetic radiation. Also don't let children sleep with their heads close to electric sockets or to a TV, even if it's behind the wall in another room.

• **Use public transport wherever possible** and/or walk or cycle your child to and from school and shops, to reduce your contribution to the pollution from car emissions and to traffic hazards, and to improve your child's and your own health.

• **Avoid heating polycarbonate plastic baby feeding bottles,** such as those made by Boots, Mothercare and Avent (see the warning on page 70).

• **Avoid using plastic teething rings,** which can pass plastic into a baby's mouth: use raw food alternatives instead (see Organic Food, Health & Nutrition Tips) or an amber teething stone necklace, which cools the mouth, as sold by Schmidt Natural Clothing.

• **Recycle and limit your waste.** Use your leftovers to make garden compost (use a simple worm composter, or consult the many useful organic gardening books now available to make your own). Use your compost to grow your own organic vegetables. Sprout alfalfa and other seeds on the windowsill using a simple glass sprouter (see Get Sprouting in Organic Food, Health & Nutrition Tips).

- **Get to know other organic and natural parents**—share information or develop a support network. There are a number of established natural parenting networks, and several supportive publications available.

- **Seek out organic restaurants, B&Bs, farms and events** for family outings (see Organic Outings in Part Two).

- **Use natural parenting methods** which suit you and your baby. Consider a home birth with appropriate professional support. Consider long-term breastfeeding and sleeping with your baby in the family bed until it feels right to move him to a cot or bed (reducing the risk of cot death and giving your child emotional and physical security), boost your child's immune system with fresh, organic foods, and use herbal tinctures and/or homeopathy rather than having potentially harmful immunisations, etc. Get the professional advice you need to support your baby's health and your own, read up on the vaccination debate (see pages 95-99), and join a group such as The Informed Parent or Natural Parent Circle for information and support.

"While for several years we bought some organic products as available in our local supermarkets, we recently decided to aim for 95% plus organic. I wish I had made more effort before, and that I had not been mean with my money. My toddler eats almost any organic food I put in front of him and really enjoys it, whereas he used to be very fussy. I used to resent time in the kitchen but now I enjoy it and feel good about feeding healthy food to my family. I hadn't realized how guilty I felt about serving up chemical food. I realized we couldn't afford not to buy real food, so economies have to come elsewhere and we cut out processed food. We have now been eating about 98% organic food for a year and it has been the most wonderful thing for us as a family. All three of us are equally enthusiastic and we go to organic events and visit gardens together, even though it's hard to stop our son Sam grabbing the tomatoes and other fruit from the bushes. It sounds a bit over-the-top, but it is as if we have completely rediscovered the taste of real, clean food, the scent of flowers, the feel of organic clothing on the skin—health and happiness. Eating organic has led us into eating raw food, and myself and my son now eat almost nothing but raw food, which has further improved our health and wellbeing. When Sam was 12-18 months old he threw tantrums, banging his head on the wall, and was unhappy for quite a lot of the time. But he has been a terrific two-year old. He used to have severe eczema, of which only a trace remains, and my husband has had no hay fever this year despite the high pollen counts."—Rosemary Langridge, South Norwood, London.

• **Avoid using dummies** as they can delay speech development and babies can become emotionally attached to them.

• **If you plan to use a child-minder or nursery,** be prepared to send in your own prepared organic food and drinks every day (including breastmilk if still breastfeeding), and ask whether they are prepared to change and bag reusable nappies. If your child is unvaccinated, ensure that carers wash their hands thoroughly before changing your baby, as viruses from other children's nappies can be passed on. A few nurseries have begun offering organic food—encourage yours to consider doing so.

• **Once you have a child you'll notice just how much food advertising is aimed at them** and how susceptible they are to it ("Take me to McDonalds!"). Restrict their viewing, and explain the difference between those products and the food you eat. Organic parents of older children report that once they understand the difference, they don't want non-organic foods.

• **If you're travelling abroad,** don't assume you can buy equivalent organic babyfood and milk at your destination—take a supply with you.

• **Use fabric shopping bags, a backpack, baskets or push-chair storage** for shopping instead of plastic carrier bags. If you're able to, cycle to the shops with your (older) baby in a child's seat, and put shopping into panniers or baskets, or invest in a small bicycle trailer. Put soiled nappies into recycled carrier bags, rather than expensive, artificially perfumed nappy sacks.

• **When crossing a busy road,** avoid going directly behind a car with a push-chair—the exhaust emissions are at child height.

• **Don't let people smoke near your baby,** or near you while you are pregnant, and avoid public places that are smoky. Cigarette smoke contains many carcinogenic chemicals which can harm your baby through passive smoking. Don't be afraid to ask people to refrain from smoking, to protect the health of your baby.

• **Act as a catalyst for other people's change**—feed them organic food and show them your organic products and explain their benefits, tell them why you do things a certain way. Don't be a bore, but instead encourage informed dialogue about important environmental, food, health, agricultural and social issues. Give them organic wine, chocolates or other organic treats for Christmas and birthdays and at other social occasions. Have an entirely organic Christmas dinner!

• **If you have pets,** look out for organic pet food such as Yarrah or Pascoe's, and avoid chemical pet treatments in the home—there are now non-toxic alternatives.

- **Put your money where your health is:** purchase organic, environmentally friendly, socially responsible (fair trade) and recycled goods and services wherever possible, and by doing so support the companies making and supplying these products and services. Choose products made from renewable resources. Tell your friends about these products and companies and let them look through the catalogues. Bank with an ethical bank such as Triodos or the Co-operative Bank, and consider accounts such as Triodos Bank's Organic Saver Account, which contributes directly to the organic industry.

- **Make time for lots of natural outdoors fun activity** such as exploring unspoilt woods and beaches, meadows and traditional countryside, and enjoy being a parent who is making a real difference to future generations.

How to convert if you already have children

It's never too late to go organic and improve your children's health. Follow the simple steps below and before you know it you'll be an organic household. If cost is keeping you back, make sure that you at least buy the basic foodstuffs organically (the ones you use the most of), such as milk, bread, breakfast cereals, coffee, tea and sugar, and that you start to buy raw organic fruit and vegetables rather than more expensive processed foods.

It's difficult to say which foodstuffs are more important to eat organically than others, as we should be concerned about all non-organic foods. However, it makes sense to substitute whatever foods you or your baby eat the most of (including commercial babyfoods), as these will be giving you both the greatest exposure to potential toxins, and to substitute foods which are thought to absorb the most chemicals, such as small grains (and therefore bread, breakfast cereals, pasta and rice). Substituting meat, egg and dairy products is vital because of the hormones, antibiotics and other drugs, the likelihood of contaminated feed, and the inhumane treatment and high level of bacterial disease in conventional livestock and poultry. Imported non-organic foods from developing countries are more likely to be contaminated with even more dangerous pesticides that are banned in the West, and to be sprayed more heavily than western crops. The many foods that have been found to contain high residues, be sprayed excessively or with banned chemicals include oranges, carrots, apples, pears, lettuces, strawberries and bananas (always wash your hands thoroughly after peeling or handling non-organic citrus or bananas due to the chemicals on the skin, and never let your baby or toddler play with them). If you eat meat, make eating organic meat, poultry and/or fish a priority: go for quality rather than quantity, which is far healthier.

Read up on all the benefits of organic foods—and the not-so-good things about chemically produced conventional or GM products—so you'll be able to hold your own in any discussions with your family and friends. Joanna Blythman's book *The Food Our Children Eat* is great for parents with fussy eaters who need practical strategies to get them into healthy eating habits.

A word of advice: don't try and force change on to your family though, as you may meet with resistance and actually harden attitudes if they are sceptical about all things organic. Instead, introduce them gently and gradually. Organic chocolate, flapjacks or ice-cream may be a good place to start for older 'sweet tooth' members of the family, although swapping from a sugary, processed conventional diet to a sugary or sweet, processed organic diet is not the answer—try and gradually steer your family towards a wholefood organic diet with plenty of fresh food by introducing a little every day. 'Feed' them the organic message casually, and with any luck you'll have a house full of happy converts quaffing organic carrot juice and munching sprouts in their organic sandwiches in no time!

• **Look for organic food in the places where you shop now**—look for the certifying symbol. Organic produce is usually grouped together, while grocery items may be alongside other products of the same type. See if there's an organic alternative for your usual purchases.

• **Begin your conversion by choosing just a few organic items** for your shopping basket such as organic yoghurt and milk, coffee, carrots, bread, cheese and perhaps chocolate or ice-cream. There will be several brands to choose from.

• **When you return to your local store**, try some of the other organic products—kiwi fruit, oranges, apples, lettuce, frozen peas, sweetcorn and pizza, peanut butter, pasta sauce and fruit juice. Look out for information on monthly or seasonal specials and new product information.

• **You could get your children actively involved** in the family's conversion by setting them a fun task for the weekly shop—to locate all the organic goods and pick ones they'd like to try.

• **Visit a leading wholefood or healthfood store** and explore the huge range of organic products on offer—organic toothpaste, shampoo, vegetable and seed oils, ready meals and essential oils. Try some! Talk to the manager or staff about particular organic products they recommend, and make the most of their organic knowledge and advice on health generally—a human touch you definitely won't get in supermarkets!

• **Consider starting a weekly home delivery** of organic fresh fruit, vegetables and even red meat, chicken and fish if you're a meat eater (or find a trusted butcher or specialty organic meat service), and groceries if the company offers them. You may wish to sample the wares of several different companies to see which gives the best value. Most nationwide home delivery companies now provide a large range of organic grocery items, including babyfood. Some healthfood stores now sell fresh produce, or can arrange a delivery for you, to be picked up from the shop each week. And if you need the food quickly, most nationwide delivery companies can deliver within 48 hours. Some services will even reuse your food boxes, taking them back when delivering your next order, and can offer books, juicers and even organic clothing.

• **Introduce juicing and sprouting** as fun family activities. Challenge your children to come up with the most tasty or original fruit or vegetable juice, or to see who can grow the best sprouts. Make sprouts appealing by including them in pitta bread, sandwiches or chapatti rolls smothered in hummous or another similar healthy sauce (see Joanna Blythman's useful recipes for older children in *The Food Our Children Eat*).

• **Get your children cooking** or helping to prepare healthy organic family meals once or twice a week. Collect favourite recipes and let them make their own concoctions. Plan an organic picnic or dinner party with family and friends.

• **When you need to buy new towels, sheets, nightwear or clothes** for the children, or other household goods such as laundry detergent, consider buying organic and environmentally friendly products, and look through the many catalogues now available to find the best value. Consult the Products Guide for the wide range of organic and environmentally friendly products now on offer.

• **Take your family on a day's outing to an organic farm,** to an organic event or to an organic restaurant, and let them see, hear and taste for themselves why organic is better.

• **If you're concerned that you or your children may have vitamin, mineral or trace element deficiencies,** you can take a sample of hair (close cut at the nape of the neck) and send this off for hair mineral analysis, and then seek professional advice to correct any deficiencies.

• **If your children have allergies,** see whether eating organic food, drinking pure filtered water and using organic or environmentally friendly products makes a difference. Many parents have found that a mainly raw organic diet has significantly reduced allergy symptoms, or even eliminated them. If not, seek professional advice, have an allergy test, or exclude possible allergens from their diet for a month at a time, and then reintroduce them one at a time to pinpoint the cause.

Organic food, health & nutrition tips

Many health experts and nutritionists agree that the following should be among your nutritional priorities for both healthy parents and a healthy baby:

- **Eat as much raw, fresh organic produce as possible**, even in winter, always choosing fresh rather than processed foods.

- **If you must cook, lightly steam, quickly stir-fry or lightly boil** wherever possible. The longer you cook food, and the higher the cooking temperature, the more important nutrients are lost. However you can actually get more nutrients out of some foods that are lightly cooked, steamed or puréed, because you break down some of the very strong plant wall fibres. Quick stir-frying (with more water or sauce than oil) is ideal.

- **Try to eat in-season,** and choose locally grown produce where possible.

- **Avoid refined foods,** including white bread and flour, white rice and pasta.

- **Cut down on or eliminate sugary foods,** watch out for too many other 'hidden' sugars in the diet, and don't add sugar to your or your baby's food.

- **Make sure you're getting enough essential fatty acids** by taking oils such as Udo's Choice or Essential Balance regularly with food (1–2 tablespoons a day for adults, 1–2 teaspoons a day for babies from weaning), or ground seed blends with food, or by eating quality oily fish (organic fish is now available in some outlets). See The Good Oils on page 86.

- **Eat a variety of foods** to ensure a variety of nutrients.

- **Build and boost your immune system** (consult a practitioner and/or books such as those by experts Patrick Holford, Michael van Straten and Leslie Kenton).

- **Ensure you're getting a regular, organic supply of protein** whatever your diet.

"I went half organic before conception, and scrutinized labels. I've fed my son almost exclusively organically since weaning. While he's developing his digestive system, the last thing I want to do is to introduce pesticides and other chemicals into his body. He was a big baby, and hasn't had the ailments other babies have had: people always remark how bonny he is and what great skin he has, which makes me feel very proud. When he was first born, people remarked how he didn't look like a new-born at all—he was very advanced and observant."— Dawn Thomas, Kent.

- **Cut down on the amount of meat and dairy products you eat**—go for quality rather than quantity, and choose lean cuts of meat rather than sausages, especially for children.

- **Eat more pulses, wholegrain rice, cereals, pasta and noodles** (for babies, from 8 months onwards).

- **Cut down (or even better, cut out) salty, fatty and fried foods.** Don't offer them to your baby, nor add salt to your or your baby's food: try seaweed condiment or soya sauce in small amounts on your meals instead.

- **Never feed a baby cocoa or chocolate,** even in organic ready-made babyfoods, as they contain the stimulant caffeine, which is addictive, and another stimulant, theobromine.

- **Don't offer your baby or child commercial sweets and snacks,** most of which are laden with additives, sugars, fats, salt, gelatin and possibly GM ingredients and transfatty acids—he doesn't need them, and will only develop bad eating habits. (Organic ones will still contain sugar, although in much lesser amounts, and organic crisps usually contain salt, although much less than regular brands.)

- **Avoid hydrogenated oils or fats** (read the label): they create transfatty acids, which are anti-nutrients and harm the body (see The Bad Oils on page 86).

- **Consider supplementing with high quality vitamins and minerals and/or food supplements,** as professionally advised.

- **Drink plenty of pure, clean water** (at least six to eight large glasses a day) and ensure you have an effective water filter system, preferably plumbed-in. Get your baby used to drinking pure water as soon as possible. (Never give your baby tap-water, as the high nitrate and other chemical and bacterial content can poison.)

- **Exclude alcohol during pre-conception, pregnancy and breastfeeding** (some reports now claim that alcohol can have worse effects on the foetus than cigarettes); quit smoking, avoid passive smoking, and avoid any other recreational drugs.

- **Similarly, avoid all artificial sweeteners and additives,** some of which have been linked with cancers and hyperactivity, and many may be made with GM ingredients.

- **Cut down on or eliminate carbonated drinks** as they are said to leach nutrients out of the body.

- **Make your own fresh juices** rather than buying processed juices in cartons or bottles, which will have fewer vitamins and enzymes.

- **Avoid tea and coffee** while pregnant or breastfeeding, and ideally during pre-conception: try herbal teas and coffee substitutes (seek advice on suitable herbal teas).

- **Cut down on or avoid tinned foods**—tin has been found in canned foods and can be inadvertently dropped into the food when cutting open the lid.

- **Cut down on or avoid drinks in cartons**—many of these are lined with aluminium foil and the aluminium can pass into the fluid.

- **Cut down on or avoid drinks in aluminium cans**—some of these drinks have been found to contain up to six times the level of aluminium compared with bottled drinks.

- **WARNING: Do not heat or boil plastic baby feeding bottles.** Most popular feeding bottles on the market are made with polycarbonate plastic. A report published by the World Wildlife Fund in 2000 highlights the dangers of some plastic feeding bottles because of exposure to Bisphenol-A (BPA), an industrial chemical used to manufacture polycarbonate and other plastic items. The level of exposure in a bottle-fed infant is less than the tolerable daily intake, but greater than the quantity found to cause effects in studies on animals. Storing, heating and sterilising plastic bottles can enable chemicals to leach into the milk. WWF are particularly concerned about younger infants, who might use their siblings' bottles which are older and have had more exposure to dishwashing and bottle brushing. Manufacturers are being asked to include advice to consumers on the label to change bottles every six months. WWF adds that "a change to safer materials is preferable".

  A scientific report by American and Italian universities published in the *International Journal of Toxicology and Industrial Health*, featured in the *Sunday Times* in August 1998, revealed that, based on animal studies, alterations in reproductive organ development could occur in the foetuses of pregnant women who consumed bisphenol-A in canned products or from foods heated in polycarbonate containers.

  Avoid any baby feeding bottles made with PVC (these are now rare).

> "The cost of organic food and other organic products has been an issue for us—we buy less food as a result. I would like to buy everything organic but price forces me to stick to bread, milk, eggs, fruit and vegetables that are available organically. I suggest looking for reduced organic food in the supermarket—there's always something affordable and if that product isn't bought by anyone they might order less organic produce for the next delivery."—Yvan Van Royen, Harrow.

Why raw?

Several health experts and nutritionists recommend eating mainly raw food—food in its natural state—for improved health. Through her many bestselling dynamic health books, Leslie Kenton has been promoting the value of eating raw, living foods for many years, showing how they are at the root of a healthy high-energy diet and provide living water full of nutrients necessary for life. Nutritionist Patrick Holford agrees, and says in his bestselling book *The Optimum Nutrition Bible* that every time you eat a combination of fresh, living foods, you are giving yourself a cocktail of essential vitamins, minerals, amino acids, antioxidants, enzymes and phytochemicals which work together to promote your health. Raw organic food, he says, is the most natural and beneficial way to take food into the body; and cooking food destroys enzymes and other beneficial nutrients. Nutritionist Antony Haynes of The Nutrition Clinic says that a good principle to apply in your daily diet is to have at least half of everything on your plate as fresh, or something that you can see growing in nature. Have you ever seen that flapjack growing in that flapjack tree, he asks?

"Babies raised on a mainly raw, unprocessed diet are much less likely to experience the unnatural cravings for refined sugar which can send their body biochemistry haywire and lead them to become physically addicted to sweet processed foods," says Karen Knowler, co-ordinator of The Fresh Network, an organisation promoting a raw food diet for both children and adults. Young children given sugary processed foods are soon at the mercy of their taste buds, and can suffer behavioural changes and blood sugar problems which could even lead to diabetes. If they only receive their sugar supply from foods in their natural state such as fresh fruit and vegetables, they will be much healthier and happier, she says. Some network members have brought their babies up on 100% raw foods, and report that the children are well behaved and content, and don't feel left out, or that they are missing out as they get older when eating with others—in fact other children usually ask to try their food, because they're not given fresh fruit or vegetables at home! Karen says that living foods will create a truly living body, and that cooked foods are effectively dead. "When you eat raw food you're taking in new enzymes which help the body do all it has to do to stay healthy and strong, but when you eat cooked food no new enzymes are introduced at all, and in fact existing enzymes already in the body are used up to process that food. In our early years these come firstly from the digestive system, but usually by the time we reach mid-life, they are having to be robbed from other parts of the body, leading to degenerative disease. Our members have found that by eating a diet consisting of 80 to 100% raw food they and their children recover from allergies, eczema, asthma and many other often serious health problems because the body is finally being supported nutritionally in order to facilitate self-healing."

Changing to even a partly raw wholefood organic diet can make you feel more alive, energized and alert. Waking up in the morning no longer becomes a chore, and you'll feel better and have more energy (a real boon for any parent), even if you're enduring broken sleep with a young baby. And if it makes a difference to you, think what it can do for your baby, who needs as many unadulterated, high quality nutrients as possible to build and develop his body. Mainly raw meals are best for baby, and there are juice, food blend recipes and raw finger foods featured in this section—compiled with the help of Penrhos School of Food and Health organic chef/nutritionist Daphne Lambert. A complete guide can be found in *The Organic Baby & Toddler Cookbook*. Try to start your baby off with good eating habits from the day you start weaning him. As Antony Haynes says, you won't be able to control what your children eat beyond a certain age, but you can while they're babies, and can set them up for better health while they're young.

General food & health tips

A number of these useful tips have been contributed by organic chef/nutritionist Daphne Lambert; Patrick Holford, founder of the Institute for Optimum Nutrition; and organic mother Helen Pitel, who lives in London.

Be prepared for your new role as family chef and nutritionist for yourself and your partner during pre-conception, pregnancy and breastfeeding, and for your new baby during weaning and beyond. It takes a little effort to prepare interesting and nutritious meals, but once you have your kitchen in order (see Useful Kitchen Equipment and Suggested Store-cupboard), have your sprouting and juicing system going, all the fresh ingredients you need and some recipes to try, you'll never look back.

• **If you can afford it, book yourself, and if possible your partner, on to an organic cookery or food course** (ideally one which covers both food and health) to learn about practical, good nutrition and how to easily make extremely healthy, mainly raw organic meals for both yourselves and your baby. Learn how to incorporate all the nutrients your bodies need, including a balance of vitamin and minerals, essential fatty acids, protein and complex carbohydrates. Learn how easy it is to include raw juices and sprouting into your daily meal planner. Green Cuisine at Penrhos Court in Herefordshire offers short courses which are both practical and informative. Regular two or three day courses throughout the year cover Organic Food and Health, Women's Health, and Mother and Baby (see *Product Listings* guide). Alternatively, or in addition, read up on some of the many excellent books and recipes by Leslie Kenton, *The Organic Café Cookbook* by Carol Charlton Jones, *The Optimum Nutrition Cookbook* and *100% Health* by Patrick Holford, *Organic Super Foods* by Michael van Straten, *The Foresight Wholefood Cookbook*, *The Organic Baby & Toddler Cookbook* by Tanyia Maxted-Frost and Daphne Lambert, and *What Should I Feed My Baby?* by Suzannah Olivier. A new recipe book, *Optimum Nutrition for Babies and Children* by Lucy Burney, could also be useful.

• **Make your own meals as often as possible,** only falling back on processed foods once in a while, if at all—especially for your baby (see 'Get sprouting' and 'Get juicing' sections.)

• **Don't be put off by the odd mis-shapen organic vegetable,** bug in the lettuce or pock-marked organic fruit. The reality of organic farming is that it produces healthy, quality food, as opposed to the cosmetically uniform, relatively tasteless chemically grown produce usually found on supermarket shelves.

• **Be prepared to eat unusual fruits and vegetables** which will turn up in your organic food boxes from time to time, such as mizuna (Japanese greens), chard, kale, Florence fennel and kohl rabi. Many box schemes offer useful recipes (or they will if you phone them), or you can invest in a good organic cookbook as suggested above.

• **Don't use a microwave oven**—there are many concerns about the effects on both your food and health of using microwaves. Cataracts, neurological injury and pacemaker dysfunction are among medical hazards associated with microwave ovens, according to *The Perils of Progress* published by Zed Books (see Publications in Part Two). The book quotes studies showing that microwaved food produces adverse changes in human blood and molecular changes in the amino acids in milk proteins of infant formula, causing toxicity.

"I have found that the genetic modification of tomatoes has produced a thicker skin. (GM) tomatoes often look a little squarer and are harder to cut. This is actually done on purpose so they can be thrown on a conveyor belt and dropped on the end of trucks without blemishing. We recently found that the pressure tolerance of these tomato skins exceeds the required bumper tolerance in new cars."—Professor John Wargo, Centre for Children's Environmental Health, Yale University, US speaking at the London Food, Children & Health Conference, 1998.

• **Keep imported foods to a minimum**—choose UK-grown and processed. Fresh foods which are available year round can come from up to 13,000 miles away, and because of the time in transit, they probably won't be high in vitamins and minerals. They also cost the earth in transport pollution. Unfortunately, during the 'hungry gap' in between UK growing and harvesting seasons, up to 70% of organic produce on sale can be imported.

• **Don't be tempted to start your baby on solids early,** as this can cause digestive problems. If your baby seems restless on your breastmilk, concentrate on improving the quality of your milk by improving your diet with more organic, mainly raw wholefoods and herbs, etc, and by resting more. Breastmilk is all he needs for the first four to six months.

• **Use ripe fruits and vegetables for your baby's food**—they are more easily digested.

• **Babyfood shouldn't be bland**—taste preferences developed in infancy pave the way for those in adulthood, so create flavoursome, interesting meals for your baby.

• **Fresh purées of single organic fruits and vegetables** (such as raw apple, raw pear or lightly cooked squash) are best for starting baby on solids. Mix such foods as carrot purée with breastmilk to begin with: this helps babies tolerate new foods, as they're used to the flavour of breastmilk.

• **Consider excluding eggs and cows' milk from your baby's diet** until he is at least a year old, to try to prevent allergies; and then consider using unpasteurized 'real' milk, which still contains important nutrients that aren't in pasteurized, homogenized or UHT (ultra heat-treated) milk. Wheat, oranges, gluten and nuts are other potential allergens whose introduction can also be delayed, especially if there are allergies in the family. However, in some parts of Europe (and more generally, for vegans) it is common to include ground nuts in weaning babyfood, for their protein and other nutrients.

• **Don't make dinnertime battle time**—if your baby refuses a new food, reintroduce it at another time, or perhaps disguise the flavour with another food if it is a vital nutrient. For example, try putting a purée on to organic rice cakes or toast (if he has teeth)—babies will often eat it (on their terms) this way.

• **Eat fresh foods as soon as possible after purchase,** freeze any you can't, and buy carrots and potatoes with mud still on them to preserve their storage life—organic foods often don't keep as long as those that have been chemically treated and/or irradiated, especially in summer. You can always freeze half a loaf of organic bread and defrost it when you need it.

• **Tofu and soya yoghurt can be slowly introduced** from eight to nine months onwards, but don't over-use them. Soya and other dairy-free organic milks (oat, almond and rice, and chemical-free pea and coconut) are also fine in small quantities from this time onward as part ingredients.

• **Limit the number of commercial baby rusks** that your baby consumes, due to their sugar content. Consider making your own. Beware of giving refined white flour products such as commercial baby breadsticks—always favour wholemeal products and ones without gluten for young babies.

• **Avoid buying foods packaged in aluminium foil.** In his book *100% Health*, Patrick Holford claims that it can adversely affect zinc levels and is associated with premature senility and memory loss. Unfortunately, a lot of leading brand organic packet cereals for babies are packaged in aluminium foil, and many juices and milks are packaged in cartons lined with foil. Choose foods packaged in cellophane or paper, or bottled, wherever possible.

• **Use grease-proof paper as food wrapping and covers** instead of plastic film or aluminium foil—the paper can be composted and will not contaminate food. Similarly, don't use aluminium foil for cooking—cook with lids instead.

• **Don't heat plastic milk baby bottles in hot or boiling water**, as it's been shown that the plastic migrates into the milk (see warning on page 70).

• **Don't cook in aluminium saucepans or plastic-coated non-stick frying pans, or use plastic spoons or spatulas for cooking** for the same reason—instead use stainless steel, iron (cast iron woks are great for cooking all kinds of quick and easy meals), ceramic, or wooden (spoons).

• **Avoid heat-sensitive plastic baby bowls and spoons** which change colour when hot (use stainless steel or ceramic ones).

• **Choose loose fruits and vegetables** rather than pre-packaged, especially where plastic is concerned, choose foods and drinks bottled in glass rather than plastic, and don't cook or heat plastic containers such as those used in ready-made meals: transfer them into baking trays or dishes.

• **Smell fruits before you buy them**—good ripe ones should have a delicious aroma.

• **Freeze left-over home-made adult meals or babyfood** in ice cube trays so you can thaw or re-heat small portions at different times. Don't freeze and then re-heat commercial babyfoods, and only keep them in the refrigerator for a maximum of 48 hours. Clean bottled jars and lids in hot soapy water immediately after use, and reuse for your own concoctions or recycle.

• **Use freezer or food bags to keep larger leftovers,** and wash and reuse them.

• **For teething babies with several teeth,** try large strips of organic dried fruit such as mango.

• **If you or your child suffers from hayfever, trying eating local honey—** it's reputed to desensitize you to local plant allergens.

• **Express excess breastmilk and freeze** in special bags (made of inert plastic) or as ice cubes, for use in foods. It's a good idea to continue this practice as long as possible to help keep your milk supply up, to be able to use your milk to add to weaning foods, and to ensure breastmilk is available for your baby if you're away from each other for several hours. You could also use inert plastic or glass containers.

• **Probiotics are recommended to re-establish friendly gut and bowel bacteria** after taking a course of antibiotics. They can also be taken during a long antibiotics course by you or your baby, or if recommended by a practitioner as part of a restorative health programme. Add to food or drink in recommended quantities—there are special baby probiotics available. Some organic 'superfood' blend supplements in powder form contain probiotics and digestive enzymes, and can be given in reduced amounts to babies and children.

• **Be wary of taking fish oils**—some regular brands have been found to contain high residues of chemicals such as PCBs, as oily fish are relatively high up the food chain, where chemicals become more concentrated.

• **Frozen vegetables can contain more vitamins than some 'fresh' produce** which has been sitting on the shelves, in transit or in storage for some time, as the frozen ones are processed soon after harvest, but on the other hand, some important nutrients are destroyed by freezing. Keep some packs of frozen organic vegetables in your freezer as a back-up— especially peas, which can be quickly brought to the boil and then puréed and cooled for your baby.

• **Equip your kitchen with useful time-saving devices** to make feeding yourself and your baby as simple and quick as possible. Stock up on essential items and staples which you can use for or add to any meal, and to avoid lots of trips to and from the shops or to save on delivery charges. Avoid anti-bacterial plastic chopping boards and other such products

"The soil is a living organism. The health of man, beast, plant and soil is one indivisible whole."—Lady Eve Balfour, who founded the Soil Association over fifty years ago.

(implicated in helping to create 'superbugs'), and glued wooden boards. See the list of stockists in the Products Guide.

• **Don't over-clean your kitchen or house,** or sterilize with chemicals: babies need to be able to develop their immune system. Washing your baby's equipment with hot, soapy water (using an environmentally friendly cleaning agent) and then rinsing in boiled water is fine.

• **When buying organic ham** (not recommended for babies, despite the fact that it is used in some babyfoods), look for joints cured without the use of sodium nitrates. Nitrates make the ham look pink (it's naturally browny beige), and although these are now permitted in organic food, they are known carcinogens. Their introduction has split opinion in the organic movement about what is an acceptable food additive (the Soil Association claims that nitrates have been allowed for food safety reasons).

• **Don't peel organic carrots**—there are important nutrients in the skin. Similarly, once your baby is a little older you can leave skins on organic fruit, and organic vegetables such as potatoes when you blend them; organic cucumber skin can also be left on for finger food, once several teeth have arrived.

Useful kitchen equipment

Blender—Thermomix stainless steel blender/cooker is ideal
Grinder (for nuts, seeds etc—can either be manual or as an attachment
    to blender or juicer)
Juicer
Glass sprouter
Steamer
Cast iron wok
Cast iron or stainless steel frying pan
Stainless steel or ceramic/glass pots & pans
Chopping board made from FSC (Forestry Stewardship Council)
    sustainable wood (use a separate one for meat, if you're meat eaters)
Water filter (ideally plumbed in, reverse osmosis)
Ice cube trays for freezing leftovers (rubber ones are now available)
Freezing containers for storing the ice cubes (such as old ice-cream
    containers)
Reusable bags or containers for freezing larger leftovers
Inert plastic or glass breastmilk containers for the fridge and freezer
Breast-pump and bottles (<u>not</u> polycarbonate plastic)
Inert shaking/mixing containers to mix up simple small feeds/liquid meals
Garlic crusher
Measuring spoons and cups

Reused glass containers for refrigerating leftovers
Stainless steel bowl and spoon for baby
Baby drinking cup

Suggestions for an organic store-cupboard

Cold-pressed oils for salads and cold meals such as sesame, sunflower,
    walnut or pumpkin
Good quality olive oil for cooking (Meridian)
Essential fatty acid oils such as Udo's Choice and Essential Balance (keep
    in the fridge and use within six weeks of opening)
Ready-made babyfood in jars and packets as back up (favour whole
    grains and products in cellophane or paper—not in aluminium foil
    or plastic)
Rice cakes—adult and baby sizes (Kallo)
Baby rusks (Kallo, Babynat)
Tahini
Pumpkin, sesame, sunflower and hemp seeds
Dried seaweed (finely milled or in pieces ready for milling or adding to
    salads whole—try Seagreens capsules or condiments, or Clearspring
    Arame and Hijiki Sea Vegetables)
Wholemeal pasta (try Meridian wholewheat) and noodles (try the
    excellent Clearspring Mitoku range of Japanese Udon, Soba and
    Ramen noodles)
Brown, long grain or wild rice
Miso soups (Clearspring or Community Foods)
Rice crackers (try salt-free Wakame or Mitoku) for dips or for a teething
    baby
Soya sauce (Tamari, Shoyu)
Muesli (such as Whole Earth)
Other breakfast cereals (try those from Dove's Farm or Whole Earth
    Wholegrain cornflakes)
Beans
Lentils (brown, green and red)
Chickpeas
Alfalfa and other sprouting seeds
Quinoa and millet
Dried herbs and spices (such as from Hambleden Herbs)
Garlic
Onions (red and white)
Ginger root (especially good for curing morning sickness)
Herbal teas (such as Hambleden Herbs' range)
Frozen vegetables (such as the Whole Earth and Waitrose ranges)

Soya (such as Provamel) and/or rice (such as Clearspring's Rice Milk), almond and hazelnut (such as Evernat), milks

Coffee substitutes such as Bambu by Bioforce

Wholemeal flour

Rye flour

Vinegar (try Hambleden Herbs or Aspalls)

Bottled water if not filtering (choose water packaged in glass rather than in plastic, or use refillable large inert plastic bottles and fill with pure water at healthfood shops)

Baked beans (such as Whole Earth's)

Nuts (whole for adults or ground for children, and unsalted and unroasted) such as almonds, walnuts, pecans, brazils, cashews and peanuts

Dried, unsulphured, non-glacé fruit such as apricots, dates, mangoes and prunes

Pasta sauces (such as the Meridian range)

Coconut (whole or unadulterated milk/cream)

Get sprouting

Sprouts are highly nutritious, fast-growing—and cheap! You can sprout organic alfalfa, chickpeas, green lentils, mung beans, sunflower seeds, watercress, wheat and many other pulses and seeds. Sprouted mung beans have three times more vitamin C than the best orange. As nutritionist Daphne Lambert says, "When babies are young, puréed sprouts are fantastic food for them, and you can even juice the sprouts. It's also interesting for children to be able to see sprouts growing and then eat them—they love it. And you've got something organic and cheap on the windowsill. I use old sweet jars—I have eight of them on the go. Make sure they're glass and not plastic."

Sprouting is extremely easy, once you have a system set up. Use a simple sprouting device such as the Eschenfelder Glass Sprouter (see Part Two), or make your own. Soak the sprouts in water in the jar overnight, drain, rinse, and then stand on a windowsill and rinse and drain several times a day. Within two to three days you'll have an edible crop! Add them to any fresh salad or meal for extra nutrition. See the Products Guide for suppliers of sprouters and seeds.

For a nutritious and simple meal for your baby, you could try a blend of raw alfalfa sprouts and ripe raw, peeled avocado, with a teaspoon of Udo's Choice. Wheatgrass (which is highly regarded for its nutritional content as a health superfood) is also easy to grow, although you can buy it as powder or in tablets if you're unable to grow it yourself. Soak the seeds overnight, sow into a seed tray with composted soil and leave them to grow with regular sprinkles of water each day—you can even grow your wheatgrass on the windowsill. To juice the grass you will need a manual or electric juicer that can handle it, as only the more 'professional' juicers can. Daphne advises that you introduce wheatgrass slowly into the diet—especially for pregnant mums and children. There are excellent books available on both the benefits of sprouting, and of wheatgrass juicing. They include Leslie Kenton's *Raw Energy* series, and Dr Ann Wigmore's *The Sprouting Book* and *The Wheatgrass Book*—available from the Fresh Network, Raw Health or Wholistic Research Company.

Get juicing

Consider making your own fresh, raw organic juices for yourself, and for baby from weaning. *Super Juice* author and naturopath Michael van Straten maintains that the nutrients in fresh juices are more easily absorbed by the body than in pill supplements. Processed juices you buy on the shelves are actually vitamin-deficient, and are often made with concentrates. Simple, single juice recipes are best for babies up to eight or nine months: carrot or beetroot (perhaps watered down, as beetroot has a strong taste) are ideal. Serve as soon as you juice, to enjoy the creamy fresh taste. Nutritionist Antony Haynes warns that even vegetable juices are extremely sweet, and that pure carrot juice can elevate blood sugar levels quickly, causing blood sugar highs and lows. He advises that you dilute both vegetable and fruit juices, to prevent blood sugar problems. Daphne Lambert agrees, and is of the opinion that one juice a day for babies is fine; for older babies from eight to nine months old you could introduce wheatgrass, alfalfa sprouts, celery tops, fennel (great for digestion), mint and spinach, and watercress juices. Try adding one or two different greens to a base of carrot as they all taste quite strong. You can also try a little capsicum as well, as this has a sweet taste when juiced. There are some excellent juicing books with recipes for adults that you can try for yourself, and adapt for babies and older children (use mild ones only): try *Super Juice*, which has recipes for healing and detoxing as well as boosting health, and the *Raw Energy* series and *Juice High* books by Leslie Kenton. See the section on juicing in *The Organic Baby & Toddler Cookbook*.

Herbs to grow & use

Fresh herbs are easy to grow from either seed or seedlings, and offer a cheap supply of tasty nutrients and flavours; they can easily be added to family meals and juices. They include parsley, chives, peppermint and rosemary. Fresh herbs can also be used as infusions or teas, although some are not recommended during pregnancy or while taking homeopathics—check with a qualified herbalist or homeopath. Daphne Lambert has found that fennel is great for nursing mothers: it increases milk flow and relieves colic in babies. She recommends raspberry leaf tea for six weeks before the birth, and finds that feverfew tones the uterus after the birth. For nappy rash, she suggests making an infusion of sweet violet. If you can't grow and make your own fresh infusions, try Hambleden Herbs' organic herbal tea bags.

Bioforce, Hambleden Herbs and Neal's Yard Remedies also make excellent organic herbal tinctures which have many uses as natural remedies (most of which are too strong for children). Daphne has found that viola from Bioforce is good for curing cradle cap (see Part Two).

Before using herbs during pregnancy or on babies, always seek professional advice from a qualified herbalist, ask herb specialists such as Bioforce or Hambleden Herbs, or consult some of the many books on the subject (e.g. *Neal's Yard Natural Remedies* and *Natural Healing for Women* by Susan Curtis and Romy Fraser; *Safe, Natural Remedies for Babies and Children* by Amanda Cochrane; or *Holistic Herbal for Mother & Baby* by Kitty Campion).

Quick organic blends for your baby

Between four and six months is the recommended time to begin weaning your baby. Once he has firstly tried a range of puréed single foods for several weeks (try raw apple or raw pear, or lightly cooked peas, or fresh raw mango, all with a little water), begin to experiment with blends of several foods—a few suggestions are printed below. Aim to get your little one on blends or mashes of your own (salt-free) healthy meals from six to eight months—you'll find that he'll be wanting to taste the food on your plate and grabbing at it in no time anyway. The blends below are listed for babies, but you'll probably find that some of them can be a useful and tasty, quick meal for you as well, or can be used as a salad or fruit dressing. Some of these blends have been known to prompt at least one baby's first word (the baby on the book cover: "Yum!"). Freeze leftovers in ice cube trays to be thawed on a day when you haven't got time or the ingredients for a fresh blend, or when you're travelling.

Raw avocado is an ideal superfood (although, like many wonderful fruits, it's not grown in Britain). It has lots of essential nutrients which can be combined with other fruits or vegetables. If making a savoury blend, mix Udo's Choice, Essential Blend or similar, and milled granules of Seagreens organic seaweed or a small amount of wheatgrass juice or powder (from 8 months) and blend with raw alfalfa sprouts; Chinese cabbage; lettuce; banana and/or other fruit such as pears and apples (ideally keep fruits separate from other vegetables, and ensure fruits are ripe—kiwi fruit can be used for older babies); watercress (from 8 months on); or spinach leaf.

Also try blending raw avocado with lightly steamed broccoli, lightly cooked peas, beans and/or courgettes.

Tahini (ground sesame seed paste) can be blended with raw or cooked vegetables to make a nutritious, protein-rich meal.

Hummous is another useful, nutritious meal enhancer which can be added to raw or cooked vegetables for older babies—and is a good way to introduce them to garlic! Ideally make your own, soaking chickpeas overnight, cooking slowly the next day and then blending with tahini, garlic, olive oil and a little lemon juice.

Seed blends are the best way to get your oils—grind them daily and add to your own raw food. A good mix is sunflower, sesame, pumpkin and hempseed—and why not add some dried seaweed while you're at it! The best healthfood stores package their own blends. Introduce them to baby's food as you would with tahini and hummous.

Rice, well cooked, can be blended with any combination of cooked, steamed or raw vegetables. Most organic packet baby rice uses refined rice—choose wholegrain instead, when your baby is over 8 months, and make your own rice meals—you can even cook and freeze rice in ice cubes for use in individual meals. Combine rice and lentils for increased protein.

Noodles and Pasta (choose wholegrain—try the Clearspring Mitoku and Meridian ranges) can be cooked, puréed and added to any savoury meal to provide complex carbohydrates. As with rice, wholegrain varieties of pasta and noodles can be given to babies from eight months on.

Lentils and most types of beans can be soaked overnight, drained and rinsed and then cooked, puréed and eaten on their own, together, or combined with rice and vegetables.

Smoothies Older babies can occasionally be given thick yoghurt smoothies (try soya, pea, sheep's or goat's milk yoghurts)—very popular for breakfast. For a substantial baby meal, combine a little organic mixed muesli (such as Eco Baby's) with a little fresh or unadulterated coconut, plain yoghurt or dairy-free alternatives, ripe banana and dried unsulphured organic fruit such as apricot and/or dates and a little soya, rice, pea, or goat's or sheep's milk. Blend until smooth.

Vegetables Use root vegetables, such as mashed parsnip, sweet potato or butternut squash purée, to provide a satisfying cooked meal for weaning babies. For older babies (from eight months) add raw wheatgrass juice and use non-smoked tofu, or give them large slices of cooked vegetable to eat themselves (with the skin on). Many other vegetables can be used together either cooked, steamed or raw—experiment! Try not to over-use tomato, as it can cause allergies.

For a calcium-rich meal, make your own almond milk (soak them overnight and then purée, or homogenize in a juicer such as the Champion) and purée this with Brussels sprouts. Kale, broccoli and watercress are also good sources of calcium.

Fruits Fresh mango purée is sheer delight on its own, mixed with a little water, and can be frozen as a luxury ice lolly for older children. The same can be done with bananas—either as a purée, or as half a banana frozen with an ice lolly stick inserted (you can also dip in honey and roll in carob or grated coconut). Try puréeing other fruits such as berries on their own or as a mixture. You can also freeze fruits such as mango and berries and then put them through a powerful homogenizing juicer such as the Champion, to make an icy treat. A useful raw breakfast blend is apple, pear and banana blended with a little apple juice.

Cashew nut cream For older babies not at risk of nut allergy, and especially for vegans, try this protein and calcium-rich meal which babies will love. Homogenize raw cashews in a Champion-style juicer, add water, Udo's or Essential Balance oil, and stir into a thick, luxurious cream. Can be enjoyed on its own, with vegetables, with hummous, or blended with banana, dried fruits, a little coconut and a little vegetable or nut milk. Can be frozen as a nutty ice-cream.

Finger foods

Getting your baby on to raw finger foods as soon as possible when weaning is great for his co-ordination and dexterity, establishing his independence, and a relief (for both baby and parent!) during teething. It's also an opportunity for your baby to become used to raw wholefoods and to establish good habits early. Peel and cut up rectangular slabs of peeled fresh, cool and juicy cucumber; circular slabs of peeled and cored pear or apple; or rectangular slices of or small unpeeled carrots (a slightly soft carrot is a great substitute for mum's finger when baby has several teeth!). You could also try rectangular slabs of celery with the strings removed, and cooked pieces of squash with the skin left on.

Wholemeal toast cut in half with any extremely hard or burnt crusts cut off, then smothered with ripe cool avocado and/or blended cucumber, can also be soothing—or at least take a little person's mind off the pain! The Village Bakery has an excellent range of specialist, tasty wholemeal breads to try (its Borodinsky rye bread with coriander, and Stamp Collection wheat-free bread are favourites of the baby on the cover), or learn how to make your own: the Village Bakery and others run bread-making courses. This is also an excellent time to try out your home-made hummous (see Useful Blends above). Organic mini rice cakes, which are light and easy for a baby to hold, can now be found in healthfood stores and elsewhere. Half a banana is a fun finger food for a baby with several established teeth to mash in his mouth, and with his fingers.

As your baby gets enough teeth to fully chew instead of just cutting, leave the skins on soft fruit and vegetables. And when he's a bit older, half peel a banana and let him hold and explore the whole fruit—he'll be fascinated and enjoy learning to peel it himself. See the suggestions for ice lollies for older children in the fruit blends section above.

Supplements—to take or not to take?

If you want to take supplements, there are now many special pre-natal and organic superfood ones to choose from for pre-conception, pregnancy, breastfeeding and beyond, and even special powdered vitamins and minerals for your baby, and chewable varieties for older children. Recommendations as to the use of these and general vitamin and mineral supplements by nutritionists, naturopaths and other health practitioners, vary greatly. Some believe they're unnecessary and an expensive waste if a good, balanced, mainly raw, organic wholefood diet, drawing on a wide range of nutrient-rich, mainly seasonal foods, is being eaten; if there's no environmental or emotional stress; and if most environmental toxins are avoided. They worry that people will see supplements as a quick fix, and use them as an excuse to neglect the quality of their daily diet, which they believe should be their first priority. Others believe that achieving this ideal diet (and stress-free and toxin-free lifestyle) is rare, and that key vitamins and mineral supplementation are vital to ensure that you're 'fully covered'. They also believe that if you're planning a baby, are pregnant, or are breastfeeding, you need the supplements in order to ensure that the baby has access to a healthy bank of the vitamins and minerals as it requires them, without draining the mother's supplies. Supplement advocates also recommend folic acid supplementation to prevent spina bifida.

While they both recommend an organic wholefood diet, the Foresight Association for the Promotion of Pre-conceptual Care and nutritionist Patrick Holford also advocate supplementation with formulated vitamins and minerals. Foresight has developed its own unique blend of vitamins and minerals to correct deficiencies and establish a health bank as part of its pre-conceptual health programme, tailoring supplement intake to the individual's needs; it warns against high dosages. Patrick is a great believer in using vitamins and minerals to correct and maintain health and boost the immune system, and his recommendations for adults include taking a daily dosage of between 1000 and 2000mg of vitamin C, as well as a multivitamin and mineral supplement. He has created the Optimum Nutrition Formula for Higher Nature for this purpose (but also recommends other brands); he advises adding extra folic acid for pregnancy. His books contain questionnaires on how to work out what supplements you need, based on your current state of health. He believes you should start babies on vitamins and minerals as soon as they are no longer breastfeeding exclusively, and that organic, high quality green superfood supplements—of which there are now several on the market—can also be good for boosting your health.

Leslie Kenton's view is that multiple deficiencies have become widespread as a result of years of eating nutritionally depleted foods, and that these deficiencies cannot easily be corrected: popping the latest multi-mineral tablet from your corner pharmacy or healthfood store won't do it. She says that nutrients in foods work in complex synergy: they affect each other through their interactions. "A balance of bio-available minerals and trace elements available from wholesome foods is infinitely more complex than vitamin fanatics would have us believe. To restore balance, once it has been disturbed, you need to return to good wholesome food—perhaps supplemented with extra green plants such as kelp, spirulina, chlorella, barley grass or alfalfa. This is a slow process taking months and even years."

Natural practitioner Kitty Campion, in her book *Holistic Herbal for Mother & Baby*, also steers away from recommending vitamin and mineral supplements. She says that fresh, locally grown organic produce cultivated in healthy fertile soil is the best way of ensuring a natural balance of nutrients in the diet. "Eating whole organic foods at the peak of their maturity will give you all the vitamins and minerals found in that food in their most vital forms, and besides this they will also be in the right dose and proportion to one another. There is no other place, besides foods, where these nutrients can be found in precisely this way."

Daphne Lambert agrees, and says she's very apprehensive about people taking supplements. "People think—oh, I need zinc, or I need vitamin E, but they all work in synergy, they have to be in balance. I do like wheatgrass as a food supplement in juice or tablet form, and would only recommend food supplements rather than vitamins and minerals on their own. I'd rather people get their nutrients from vegetables than from supplements. Prospective parents and pregnant mothers especially should get as many green, leafy vegetables as possible. It's the chlorophyll in the green leaves that is so important: chlorophyll helps protect against cancer and certain forms of radiation; kills germs and acts as a wound healer; and it's a good source of calcium. It's only the chlorophyll in green leaves that is going to give you calcium in the right balance with magnesium—milk products can't. I'd never advise parents to give babies vitamin and mineral supplements—I can't believe that if you're eating a really good diet, you're breastfeeding for a year and puréeing or chopping up your own healthy food, that he's not getting everything he needs. There's a danger of overload on their immature kidneys and liver." Daphne advises people that if they are going to take food supplements, they should introduce them slowly, especially when pregnant or breastfeeding.

Food supplements can give a natural boost during times of stress or increased activity—try a glass of carrot juice mixed with a teaspoonful of wheatgrass juice, spirulina, etc. See also the section on Food Supplements in the Products Guide.

The good oils

While they differ on vitamin and mineral supplements, most practitioners and experts do agree on the importance of good oils, and advise taking essential fatty acids in either cold-pressed oils or ground seeds—especially if you're not eating oily fish. The Omega 3 and Omega 6 oils are said to be essential for good health, especially important for conceiving, and for a baby's proper brain and eye development. Dyslexia in boys can reportedly be prevented by the mother taking good oil supplements during pregnancy and breastfeeding, and studies have shown that breastfed babies have achieved better intellectual performance than bottle-fed babies due to the increased amounts of essential fatty acids in breastmilk.

The best sources include Udo's Choice and Essential Balance, or freshly milled or ground flaxseed and 'power blend' seeds such as sunflower, sesame, pumpkin, flaxseed and hempseed, which you can buy in bulk. The recommendations range from one tablespoon of oil a day to

---

The bad oils

Undesirable hydrogenated oils and fats, which are found in many non-organic processed foods, are created by reacting oils with hydrogen gas at high temperatures, using a catalyst such as nickel—which can contain aluminium. This process changes the beneficial essential fatty acids into health damaging transfatty acids, which can be found in margarines and shortenings made with hydrogenated oils or fats, and which interfere with essential fatty acid functions in the body which are necessary for health. A high intake of transfatty acids has been found to make us more susceptible to disease and allergies. Fats expert Udo Eramus claims in his book *Fats that Heal, Fats that Kill*, that transfatty acids have detrimental effects on our cardiovascular system, immune system, reproductive system, energy metabolism, fat and essential fatty acid metabolism, liver function and cell membranes.

three—taken once in the morning, at lunch, and with the evening meal. There are new organic products coming on to the market all the time.

"I recommend Udo's Choice oil to everyone," says Daphne. "I put it neat on salads. It's extremely important for someone trying to conceive—it's the balance of essential oils which is critical. The best way to get your essential fatty acids is to grind up the seeds fresh daily and put them on your food. Use a mixture of sesame, sunflower, pumpkin, flax and hemp—put a teaspoon on your baby's food each day from nine months of age. Otherwise use a teaspoon of Udo's, but never heat the oil—it's air, light and heat which destroys it."

Daphne says that with her children she sieved the seeds after grinding them until the children were a year old, to produce a lovely fine flour. If you haven't taken these oils before you'll find that after taking them regularly for a while, your skin and hair condition will improve quite dramatically. Leslie Kenton even recommends covering yourself with these oils after a bath and letting them soak into the skin. You could also use them as a natural moisturiser on your face. For more in-depth information on essential fatty acids, consult books such as *The Optimum Nutrition Bible* by Patrick Holford, or *Fats that Heal, Fats that Kill* by Udo Eramus.

Other food tips

• **Get your baby used to drinking out of a feeding cup** or beaker early—this is good for his co-ordination and independence, and said to be better for his teeth than a bottle. Encourage your baby to take both water and fresh juices this way—hold the beaker for him initially and then simply offer it to him again and again to hold and drink himself. Use a straw to tempt a fussy drinker.

• **Consider green top unpasteurized 'real' milk** from a trusted organic supplier if your family eats and drinks dairy products. Pasteurized milk has been heated and essential vitamins and minerals are destroyed, whereas unpasteurized has all the enzymes and other nutrients intact. Otherwise buy organic milk which is not homogenized—it's easier to digest in its natural state.

• **You can take garlic and echinacea if you're breastfeeding** to boost both of your immune systems, if you feel susceptible to colds, instead of vaccinations, and as a natural antibiotic. Add a little garlic to freshly juiced carrot juice for yourself and your older baby.

• **Papaya tablets** (or even better, the real fruit) are recommended by nutritionist Daphne Lambert for indigestion during pregnancy. Freshly grated ginger in boiled water as a tea is great as a morning sickness cure.

Alternative health tips

• **Consider buying a homeopathic first aid kit** and learning how to use it; learn about the Bach flower remedies and keep a Rescue Remedy bottle and cream handy; learn about herbal tinctures and essential oils.

• **Teething:** as well as using a homeopathic or herbal remedy (Avena Sativa from Bioforce, chamomilla powder or crystals, or the Teetha brand by Nelsons which contains chamomilla), natural finger foods (see earlier section) cool the gums and satisfy the urge to bite and rub gums.

• **A remedy store-cupboard of aloe vera, echinacea and lavender essential oil** is recommended by Daphne Lambert.

• **You can sprinkle a few drops of lavender oil in the bath** if you and your baby have been swimming in chlorine; rub a natural oil all over baby before going in the pool water.

• **You don't need to bathe children every day**—it dries out the skin's natural oils, as do most soaps. Even organic and environmentally friendly bodycare toiletries should be used sparingly on your baby—treat them as luxury rather than necessity.

• **Aloe vera has many healing properties** and can be used to treat minor burns or scrapes—fresh from the plant is best, and cheapest.

• **Cool, juicy cucumber or aloe vera** (ideally cut fresh from the plant) are ideal for hot little faces in summer.

• **Learn and use some simple baby massage strokes.** You could use an organic mild vegetable or nut oil base, and add a drop of essential lavender oil or similar.

• **Consider taking your baby to a cranial osteopath.** Many organic mothers swear by them, and according to Daphne Lambert, seven out of ten babies are born with something out of line. "If a cranial osteopath works on a young baby, it can stop problems later in life such as bad periods or bad backs—I can't understand why the government doesn't support it, when you consider the time off work for things like bad backs. If you have a miserable, crying baby, a cranial osteopath can do wonders. I have known a cranial osteopath to turn a miserable crying baby into a happy content baby in just two visits!"

• **You don't need expensive chemical wipes for your baby's bottom:** use soft reusable organic muslin cotton, cotton pads or kitchen towels, etc, and water (which will need to be changed regularly). When travelling, or for home, you can use disposable, environmentally friendly wipes such as those made by Tushies. Alternatively, soak cotton wool in pure water and a few drops of an essential oil such as lavender, and carry in a suitable container.

- **Carry a wet flannel or muslin square** in a bag or container for sticky fingers when out.

- **Avoid using products such as talcum powder** (dangerous if inhaled), petroleum-based oils and rubs, and other additive-laden, chemical-based 'mother and baby' toiletries either given to you in Bounty bags or in other mother and baby promotions, or marketed in magazines. The actual toiletries you need for baby should be few, and either organic or environmentally friendly—consult the Bodycare section in the Products Guide. Be especially careful about creams or oils used on your breasts and nipples as your baby will ingest these—use organic vegetable, seed or nut oils only. If you want to use baby powder, look for powder made with organic cornflour, or make your own. Also avoid regular sunscreen brands (natural alternatives are just starting to come on to the market).

- **Consider carrying arnica tablets and cream** (or similar, such as Rescue Remedy) with you when out and about: children's accidents and bumps often happen you're away from home. Prompt use of arnica will always relieve pain and reduce bruising.

- **When you're stressed, try putting a few drops of Rescue Remedy** into a glass of water or straight on to your tongue. You can also use it for children—put drops into warm water and let the alcohol evaporate first.

- **You can use calendula ointment or cream** on small cuts and grazes.

The organic food shopper's dilemma

Should you buy from your local independent healthfood store, local farm shop (small family businesses) or supermarket/superstore (big business)? From a local or regional box scheme, farmers' market or national delivery company? Or a combination of these? It's the dilemma of many organic shoppers now that there is increased choice—who to support and why? Or does it simply come down to what's convenient for you? Now that there are so many options for shoppers about where to buy their organic food, perhaps it's time to consider which options have better or worse environmental and social impacts. After all, what's the point of buying organic food 'for all the right reasons' if it's actually contributing to the closure of your local family-owned and run healthfood store, loss of local jobs, increased car pollution in your area and nationally, and increased packaging production and waste and litter? Shopping locally, and buying locally produced food are promoted as being the best options by organic and environmental organisations—but do these options best suit you? The following are a few considerations.

Supermarkets

The case for

If you support supermarkets, they'll stock more (and a greater variety of) organic produce and groceries, support more organic farmers, and help switch many more consumers on to buying organic. With the buying power of supermarkets, greater areas of land are influenced around the world—if organic products are demanded by supermarket shoppers, much more of that land will go organic. By virtue of their sheer size and number of customers, supermarkets have indisputably helped the industry to grow rapidly, and raised consumer awareness by their higher profile of organic foods and promotions in-store. They have also helped to remove the 'fringe' image of organic. Some have even undertaken national TV and press advertising promoting organic food. Sainsbury's (Organic Supermarket of the Year for 2002) now sells well over 1000 organic lines in its larger stores, and Robert Duxbury, the company's Technical Manager for Organic Foods, sits on the board of the UK Register of Organic Food Standards. A survey by Mintel of 125,000 Sainsbury customers in 2002 showed that 5% of customers buy 38% of organic produce, and that the organic food market is now worth £1bn. Tesco has helped fund the UK's first organic department (at the University of Newcastle-upon-Tyne). Both Sainsburys and Waitrose offer technical support for converting suppliers. Waitrose offers over 1,400 organic lines and has over 10% of the organic market.

Most supermarkets are continually increasing their organic ranges, bringing out own-label organic ranges including convenient ready-made meals, and the major ones are involved in sponsorship of organic industry events and activities (such as the Soil Association's annual conferences), which it would otherwise be difficult for the industry to finance. In fact these major supermarkets are publicly competing for organic supremacy and the consumer's organic pound. By heavily supporting organic food and farming, supermarkets put themselves in the firing line over GM foods, and were forced to address the issue.

Supermarket prices are generally lower than local stores (in fact they can sometimes stock products at prices at which these stores can only buy them wholesale), and there are often promotions and items reduced for quick sale. Several supermarkets now home deliver for a small fee, some of them free for larger orders, or if you're an account holder. Larger branches of supermarkets will usually have a good range of organic foods, including dairy and meats, so you can do all or most of your organic shopping under one roof.

The case against

Supermarkets take business away from local stores and the high street (with their buying power they undercut local store prices, and in some cases put local shops out of business), and have changed the landscape and our lifestyles by creating huge superstores with massive parking lots out of town so that people are forced to use their cars. As well as often having a monopoly on local food supply, they can also have a monopoly on job supply and employment terms—so locals may have no other option but to work for them in jobs which may offer little job satisfaction. Food bound for supermarket shelves can clock up many food miles—not only does a lot of it come from overseas, but it can then travel to central depots, then to other depots for packaging, storage (organic potatoes, apples and pears) or ripening (organic bananas only), and then to regional depots for distribution, creating traffic congestion and air pollution, and reducing the freshness of the food. Supermarket gradeouts—the rejected produce which doesn't conform to their specifications—are being sold in box schemes and retail outlets, endangering the livelihood of some small organic producers, whose first grade produce is then not taken by these businesses as it is more expensive. Supermarkets are increasingly controlling the whole supply chain, creating problems for independent businesses, which may not be able to get an organic supply of a particular food as the supermarkets have taken most or all of it. Supermarkets use more packaging—especially polluting plastic—on their fresh produce, and create waste with plastic carrier bags and other product packaging.

There are many concerns about just how ethical supermarkets really are: with such a large share of the market they have been accused of driving down prices they pay to (conventional) farmers while their retail prices remain relatively high, and bullying suppliers. And if you use a loyalty card, just where is the information about you and your purchases going? It's thought that loyalty cards trick shoppers into believing they are getting a good deal and saving money, when product prices are actually higher than they need to be to begin with. There is often little satisfying personal contact with supermarket staff, and they rarely know their products (especially organic ones)—and sometimes even where to find them! Congestion at the tills is another drawback. Your money may go towards paying for some local staff, but most of the money you spend goes out of your area. They have been accused of only taking organic food on board—and making other changes to their practices—as a response to immense customer pressure and lobbying, rather than setting high standards and promoting healthier food and higher standards of food processing and production in the first instance. And if you live near a small branch, there may only be a limited range of organic foods available, so you may also need to shop elsewhere to get all your organic shopping. You will certainly need to shop elsewhere if you want high quality food supplements, or vitamin and mineral supplements, and organic or environmentally friendly bodycare and toiletries.

Shopping locally, at an independent healthfood store, using local & regional box schemes, or farmers' markets

The case for

The people who run local independent healthfood stores have usually been supporting organic food and farming for many years (often with little renumeration), long before the supermarkets decided that selling organic foods was viable and profitable. As stores are close by, travelling is reduced, so shopping there is environmentally friendly. With box schemes and farmers' markets, the produce is delivered by one vehicle directly from the farm or wholesaler to a market or to customers' homes, with little resulting traffic congestion and air pollution. The use of public transport is generally encouraged by the shops, and home delivery is ideal for some young families, the elderly or handicapped. The food may still clock up many food miles in some instances (but not as many as supermarket food) as there may still be imported food, and even UK-grown fresh organic food can travel to central depots before being dispatched. The obvious exception is the local box scheme which can deliver direct from the farm, or the farmers' market—giving you exceptionally fresh, seasonal food, just harvested (or slaughtered) and also direct from the farm, which is best for your body. Or you can visit the farm shop, which will also have the latest harvest and fresh dairy and meat. You'll also be able to get local specialities, such as special cheeses, breads and unpasteurized 'green top' milk. Each box scheme order may be somewhat of a surprise, and require that you expand your culinary talents as new and unusual foods appear in your order; however some now offer a 'pick and choose' list. You can also get fresh produce from healthfood stores—but this is not often locally produced. Packaging is reduced as stores and box schemes leave their organic produce unwrapped, and use recyclable paper bags and open cardboard boxes (some companies will recycle your boxes with your next order).

You can get satisfying personal contact with store, farm shop or box scheme staff. You'll often be able to get a wide range of useful products at healthfood stores, such as organic food supplements, vitamin and mineral supplements, organic or environmentally bodycare and toiletries. Store staff are often trained to be able to advise on these—you'll often find they use these themselves and can give you a personal testimonial, or the experience of their other customers. You may be very lucky and have a local healthfood store run by someone who is totally committed to health and organics, who knows about his or her products, nutrition and health issues, and organizes food and health lectures by experts, organic product promotions and consultations in-store by alternative practitioners. If you do, you may well be able to do all your organic shopping under one roof and arrange to buy some products in bulk (some stores will offer you up to 10% discount on bulk orders).

The money you spend helps keep the local store, farm shop or box scheme going, and contributes directly to your local economy, providing local jobs and feeding and clothing local children and adults. Supporting local organic production also makes your local environment cleaner and less polluted. Your best option may be a combination of both local store/farm shop and local box scheme/farmers' market, although some regional schemes (especially those serving Greater London) are continually increasing their grocery lines. Farmers' markets allow consumers to deal direct with the farm and to get a wide choice, as many different farms bring their produce. Box scheme prices often include delivery.

The case against

Unless you have a very good independent local healthfood store or farm shop, chances are that the range of organic foods will be limited, and you will have to also shop elsewhere to stock up completely. Stores may not get deliveries from wholesalers more often than fortnightly or monthly, so your favourite items may not always be available. Most are also vegetarian, so you may have to buy your organic meat and fish elsewhere, if you eat them. Local box schemes and farm shops may only sell their own seasonal produce—so you may not be able to get items you want at all, or all year round as they simply aren't grown locally all year round. You may also find that your scheme or farm doesn't use imported produce—so many fruits will be missing (this will definitely apply to farmers' markets). You may also not get a choice as to what you get each time, as most small schemes have set boxes of whatever is available. Many local or regional box schemes only deliver once or twice a week, which can mean you fall short at critical times! Farmers' markets are generally weekly, fortnightly or monthly and are completely seasonal.

Using national home delivery companies

The case for

These companies offer a very convenient, rapid delivery service by mail order catalogue: most are accessible seven days a week by phone or email, and you can also see the catalogue and place your order over the internet via a secure system (ideal for deaf people who aren't able to converse by phone). You may have a greater choice of what goes into your boxes than with many smaller schemes, and can often order by the kilo. Most deliver several days a week at one to two days' notice, and the range of foods is often greater than for local or regional schemes—some national companies have hundreds, and even thousands, of fresh produce items and groceries. They also often carry many grocery items including cakes, bread, wines, fish, meats, dairy foods and babyfoods—one company, Simply Organic, also offers sprouters, worm composters, juicers, health books and even organic clothes (it also prints its catalogue in Braille). Packaging is usually sparingly used, and produce comes in recyclable paper bags and sealed cardboard boxes. Many new products are usually added each quarter; some companies offer discounts to standing order customers and seasonal discounts on certain products.

The case against
They are generally more expensive than local or regional schemes, and you will often be charged a separate delivery fee (about £5). There can also be more middlemen in the food chain from the gate to the plate than with a local or regional scheme.

Conclusion

Only you can decide which options best suit you and your family, and you may very well use all of them at some time—especially as supplies of organic foods fluctuate with all retailers, and you will sometimes have to shop around to get what you want—or you may just want to try them all! The important thing is to shop organic whenever and wherever possible, to support existing UK organic farmers, and encourage more UK farmers to convert. Ideally, favour your local options for the sake of your local community—your independent healthfood retailer, farm shop, local or regional box schemes.

The vaccination vs healthy immune system debate

There's often immense pressure put on parents—especially the mother—to submit their baby to 'the doctors who know best' to have all vaccinations exactly when they say they're 'needed'—beginning from just two months of age. But you have the right to choose whether or not your baby is vaccinated. In recent years GPs have been benefiting to the tune of up to £2,865 for having a high percentage of immunized babies on their books, but since the controversy over the MMR triple jab, doctors want a new payment system. Pressure for vaccination also comes from health visitors, government TV commercials and literature, the media, other mothers, friends, in-laws and perhaps even your own mother. (In fact the first pressure for medical intervention starts at your baby's birth—when doctors and midwives want to give all newborns vitamin K by needle or mouth in case of a rare blood clotting disorder, but you can say no. Some supplements contain vitamin K, which will be passed on to your baby in the womb and through breastfeeding.)

Concerns about vaccination come from natural therapists, enlightened doctors and parents who have seen the side effects of vaccines in their own patients or children, and parents keen to avoid any unnecessary health problems in their babies. Safety concerns include the short- and long-term health effects of injecting babies as young as two months old—whose immune systems and organs are immature, vulnerable and still developing—with multiple live diseases straight into their bloodstream. This is not how anyone would naturally come into contact with one disease, let alone up to four at once—plus the polio vaccine, which is given by mouth. In all, fifteen disease antigens are given to babies between the ages of two to four months: some injected in four-in-one combined vaccines, and some given singly by mouth.

Possible side effects acknowledged by doctors in the government's promotional literature include fevers, rashes, allergic reactions, lumps, redness or swelling, and febrile convulsions. The Health Education Authority's *Guide to Childhood Immunisation* states that it is normal for your baby to be miserable within 48 hours of the DTP-Hib vaccine (given at two, three, and then again at four months along with the polio vaccine), and that occasionally children do have a bad reaction to the MMR vaccine: about one in a thousand will have a fit. There are many parents of autistic and impaired children who claim that vaccinations such as the MMR were to blame for their child's sudden deterioration, and Crohn's disease, cot death and asthma are now being linked to immunisations. The real reason behind starting the vaccination programme so early—at two months old—is reportedly due to doctors' belief that the mothers will be more inclined to bring their babies in for injections while they're still used to a medical routine so soon after the pregnancy and birth, rather than being scheduled out of consideration for the baby's health. And why are boys immunized against rubella? Official government literature tells us that the disease is only of concern to pregnant women (who can be immunized as children, and be given a booster as adults if necessary) as it can harm their unborn baby; otherwise it is 'very mild and isn't likely to cause your child any problems'.

There are also doubts over the efficiency of the vaccines. Critics claim that many disease outbreaks actually originate in vaccinated children, and that lowered immunity to diseases, conditions such as asthma, allergies, and perhaps even cot death can follow vaccination. Vegans and vegetarians are also concerned about animal tissues and products possibly used in vaccine production, such as egg, chick embryos, human foetal cells and monkey kidney tissue. It is also believed that aluminium, mercury, formaldehyde, hydrochloric acid and gelatin are among ingredients used in vaccines. With the new Meningitis C vaccine introduced into the UK in October 1999, and talk of vaccines for chicken-pox and cervical cancer, the number administered to each baby is rising—along with the load on their fragile immune systems.

But there are alternatives. Instead of faithfully following the herd on the mass vaccination programme from two months, you can instead choose to let your baby's immune system develop and strengthen further by starting vaccinations later, say from six months (in line with other countries such as Japan) or from one year. You can pick and choose the injections you wish your child to have, rather than subjecting them to all of them.

Alternatively, you can build your child's healthy immune system by eating a mainly fresh organic wholefood diet, including as much fun play and outdoor physical activity as possible, and using immune system-boosting organic 'superfoods' and food supplements, herbs and other natural remedies. You can let them catch childhood diseases such as measles and chicken-pox naturally (even deliberately), and use homeopathic and other natural medicines if and when they arise. There are also homeopathic versions of vaccinations (called nosodes), which are considered much safer and kinder to babies' health than conventional vaccines.

There are many informative books on the subject, such as *The Vaccination Bible* by Lynne McTaggart, and the handbook *Everything there is to know about vaccination: a guide to every major UK vaccine* from the Vaccination Awareness Network UK (VANUK), which parents should read to get a full perspective before making a final decision or giving in to peer pressure. Contact What Doctor's Don't Tell You, The Informed Parent, VANUK or other support groups (listed under Parenting and Environmental Organisations). They can provide you with regular informative newsletters, cuttings of research and articles on the subject, relevant books and even tapes of conferences debating the value, safety and effectiveness of vaccinations; you could also read books such as *A Handbook of Homeopathic Alternatives to Immunisation* by Susan Curtis, and *Safe, Natural Remedies for Babies and Children* by Amanda Cochrane, which contains remedies for babies being vaccinated and those going vaccine-free.

Also get advice from a qualified and trusted nutritionist, naturopath and/or homeopath who can support your baby's immune system through whichever course of action you decide upon, and make a point of asking your GP and health visitor for information on the jabs. My own GP said

Enjoyed by thousands of babies....

A high quality formula based on first choice ingredients from organic agriculture.

babynat
6 months
Advanced Nutritional Composition
sucrose free
gluten free
Organic follow-on milk

SOIL ASSOCIATION · ORGANIC STANDARD ·

Advanced **N**utritional **C**omposition

Organic pioneer, Babynat proposes a complete range of babyfoods*.
Whatever your favourite, Babynat assures that all of the recipes are carefully selected and prepared with quality ingredients, ensuring excellent taste, and a healthy balance of nutritional value.
Babynat, a safer choice for happy babies.

* fruit and vegetable jars, jars with meat or pasta, biscuits, juices, herb infusions and rice desserts.

babynat®

quality organic baby food.

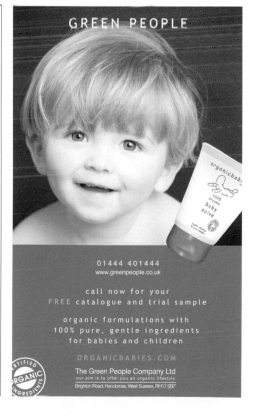

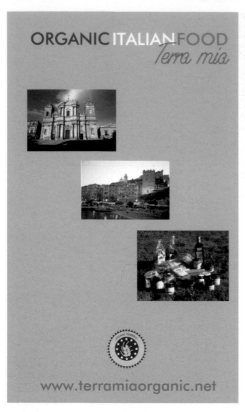
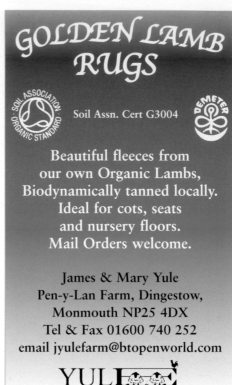

CARING

FOR

YOU

NATURALLY

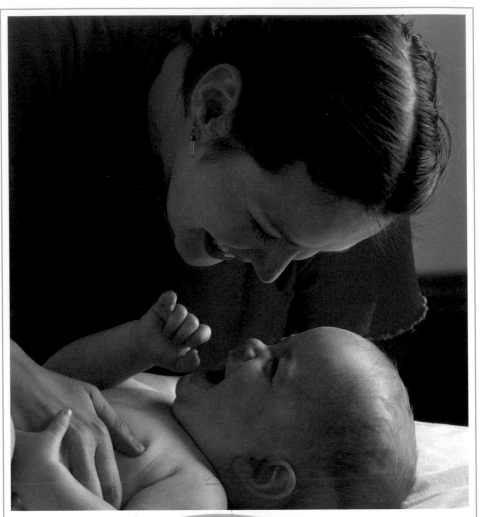

345 Upper St
London, N1

Free trial pack of
Tushies baby
wipes with your
first order.

Complete range of natural products for your new baby and you
Great choice of organic cotton clothing, bedding, washable nappies, eco disposables and much more.

Call 0870 240 6894 for a free catalogue.
Shop online www.greenbaby.co.uk Quote: GBK03

there wasn't any information (other than the standard promotional leaflet), that I should just get them done, and done at the right times! Was it my child's health he was concerned about, or his bonus for achieving a high number of vaccinated babies?

What pro-organic experts say about vaccination

Daphne Lambert, Nutritionist, Penrhos School of Food and Health: "I certainly don't think you should vaccinate babies. Their immune systems are only just developing, and the side effects of overloading the immune system can be really detrimental. I had my eldest son Adam innoculated right at the beginning, and the first jab really affected him and that's why I didn't have any more done. It had an adverse effect on him and we're just so lucky it wasn't serious. He was vulnerable in many ways when he was younger and I'm sure this was related to vaccination. My youngest son Ben, who had no vaccinations at all, has strong health and vitality. It's crazy to give so many vaccines at once, and now they're talking about vaccination against cancer for all children!

However, you can't only decide not to vaccinate: you have to boost the immune system, and do this by eating organic and fresh foods such as green leafy vegetables and garlic. It's important for you to be responsible for your health and the health of your baby. If you eat the right foods while you are breastfeeding you are passing the correct nutrients on. Once your baby is weaned you must give them the right foods (fresh and organic) to boost their immune system so they are able to fight any infections.

The main disadvantage of eating non-organic produce (aside from pesticide residues) is the adverse mineral balance it gives you and your baby. Because chemical farmers throw on loads of nitrogen, potash and phosphate to encourage growth (which makes all these huge plants) it actually causes an extremely low concentration in these plants of zinc, chromium and selenium, which are some of the really important minerals. So not only are they in low concentration in the plants, but they're not put back into the earth, so they get to be less and less, and that links in strongly with something like vaccination, because zinc is so important in the immune function. If you're not going to vaccinate you've got to make sure you're getting zinc.

The only way to be sure of a steady supply of organic zinc is by eating fresh organic vegetables. Garlic is great for boosting the immune system, for using instead of vaccinations and as a natural antibiotic."

Antony Haynes, ION/Foresight nutritionist, The Nutrition Clinic: "Vaccination is a big question mark. I personally know of cases—and there are documented cases—where individuals have autistic children after the MMR. I encourage parents to have broad reading on the subject—*Townsend Letters for Patients and Doctors, What Doctors Don't Tell You, The Informed Parent*, etc. When I have kids I will have homeopathic versions and maybe vaccinate for one or two life-threatening diseases, but not measles—we used to have measles parties! Take care of the child's immunity prior to and after vaccination with the help of a qualified practitioner.

I'm very sceptical about the benefits of some vaccinations. I'd like to see the proof that they work and learn of potential side effects. Everyone should read books such as *The Vaccination Bible* and then they will have a more informed opinion."

Patrick Holford: "Eating organic food strengthens your immunity and that is a valid approach to dealing with childhood infectious diseases. Generally a child with a good immune system is less likely to have severe symptoms to childhood infections."

Lynne McTaggart, author of *The Vaccination Bible* and editor of *What Doctors Don't Tell You*: "It is with vaccines that the brave-new-world technocrats of medicine have lost all reason about disease and its prevention. So steadfast is this faith in the rightness of their cause that it prevents doctors from acknowledging clear factual evidence demonstrating the dangers and ineffectiveness of certain vaccines, or even cases of a disease in children who have been vaccinated against it. It also turns otherwise reasonable doctors or scientists into bullies and hysterics, shouting down dissenters, using emotional blackmail to bully parents into submission, and resorting to emotive appeals, rather than common sense or fact, to argue their point of view.

When the Department of Health commissioned a report on the whooping cough vaccine by Professor Gordon Stewart (now an advisor to the World Health Organisation, and who has long studied the vaccine) and his studies showed the risks of the vaccine outweighed the benefits, the DHSS referred the report to the Committee on the Safety of Medicines, which chose not to act on it. Amid the rush to 'conquer' every possible disease, in which entire reputations rest on defending vaccination at all costs, no one is pausing to examine the possible long-term effects of pumping different antigens into the immature immune systems of a generation of babies. Epidemiologists have never investigated whether there is an upper limit to the number of jabs a baby can tolerate, after which all sorts of subtle damage (asthma, learning disabilities, hyperactivity or chronic earache, for instance) come into play.

At the heart of the logic behind vaccination is the theory of herd immunity: that is, if enough people get vaccinated against a certain disease, it will eventually disappear. Besides an element of wishful thinking in the face of highly complex organisms such as viruses, which constantly mutate and change, the problem with this line of reasoning, of course, is its tyrannical approach: eliminating a disease is more important, in the eyes of medicine, than your child's health, which might be damaged from a vaccine, or your right to decide what is best for your family. Decide against vaccination for your child and you are considered not only an irresponsible parent but an irresponsible citizen of your community and even the world."

"I first chose reusable nappies for financial reasons because I was studying for my PhD and wanted to keep costs down. My other consideration was for the comfort of my baby: if you had to be in a nappy for 24 hours a day, seven days a week for two or more years, I believe you would choose comfy cloth rather than plastic and paper. I've talked to incontinent adults, and they say that disposables are the pits, and are hot, sweaty, itchy and uncomfortable—but our babies can't tell us this. I also wanted to teach my baby that we take care of our waste, we just don't throw away—and he used to pull his nappies out of the laundry basket for me from six months, and now he sees his baby brother wearing them. He's very clued up and wonders why people use disposables. Also when they're learning to walk, falling back onto a cloth nappy with padding is much softer for them than crashing straight onto their coccyx with a disposable."—Gina Purrmann, The Real Nappy Association

Organic Products Guide

A BRIEF SHOPPER'S GUIDE

This guide has been researched and compiled to give prospective, expectant and new (or second time around) parents useful information on the wide choice of relevant organic products now available. Over 750 are reviewed, and there are also many useful contacts, services, publications etc. Current prices (as of January 2003) are given to enable readers to get an idea of costs; needless to say, these are subject to change, and different retailers will offer the same products at different prices. Many manufacturers and companies now offer mail-order services and further details about these companies, including how to order, postage & packing charges, etc, are either listed in the text or in the Leading Mail Order Suppliers section on pages 164–167.

Labelling

Look for the symbol: organic certification will be on the product, on a label attached, or on the packaging. If there's no symbol (importing companies will often remove original labelling and replace it with their own, which may exclude the product's certifier) and the product is being marketed as organic (which means it must have 95% or more organic ingredients), you can ask who certifies it—and ask to see a copy of the certificate if necessary. Common symbols on products featured in the guide such as babyfood, clothing, food supplements and some bodycare toiletries include the following:

The following symbols are used for food products only:

If a product is simply made with organic ingredients (products containing 70% or more organic ingredients can use the 'organic' term in their ingredients list), you also have the right to ask who certifies that the particular ingredients are organic.

UK organic standards for 'health and beauty' products are currently being created by the Soil Association, and are already being applied by leading UK brands in this area such as The Green People Co, Hambleden Herbs and Neal's Yard Remedies. Some UK brands have indicated they will not be seeking certification for their products.

Beware of products being falsely marketed as organic: make sure that they are certified as this is your only guarantee. Organic certification of products ensures that high standards of product, production and processing are being met. False use of the trusted and legal term 'organic' misleads, and could even endanger some sensitive consumers who rely on the organic status. It also undermines the organic industry, and puts real organic products and companies at a disadvantage, so if a product's label says it's organic, don't be afraid to ask who certifies it, and ask for proof if necessary. (If they can't satisfy your request, let your local Trading Standards Department or an organic certifier such as the Soil Association know.)

Non-organic cotton accounts for nearly 25% of total global insecticide use, so it's welcome news that in February 2003 the Soil Association introduced certification of organic textile products, covering wool, hemp, linen, cotton, skins and leather products. Such products can now carry the Soil Association symbol.

There is also the Oeko-Tex 100 Standard ('Confidence in textiles—passed for harmful substances') logo means that the clothing, from raw product to finished garment, has been tested to guarantee that no chemical residues are in the fabric.

Beware of use of the term 'natural': this has become virtually meaningless, as you only need to have a very small percentage of natural ingredients in order to label a product as such, and many so-called natural products actually contain many harmful chemicals. Bona fide organic companies use the term correctly; however 'natural' is also used by clothing companies to describe an unbleached, ecru cream colour cloth.

General product buying advice

Allow between two weeks to one month for delivery of some goods such as bedding and clothing—check when you order. Some companies can arrange swift delivery for a small extra charge, and delivery is usually between five and ten days.

Organic companies generally don't sell their lists to other companies, so you won't be bombarded by junk mail. You are likely, however, to receive regular updates on the company's products. Secure ordering facilities are available on many internet sites, but check with the company first to make sure.

The amount and diversity of organic products available in healthfood stores varies enormously. Independent healthfood, wholefood and organic stores, and organic supermarkets, carry the most (fresh, groceries, supplements, bodycare, etc) and their staff will also be knowledgeable about food and health issues, and the products they stock, so they should be able to answer any questions you may have. Such stores will mainly be run and staffed by people who care about the environment, health and the food we eat, and the importance of organic and GM-free: the stores' products and service will reflect this. Support your local independent health store!

Several organic and 'green' baby shops have opened around the country. Most of these also do mail order and will be indicated as such. Shops can be located through the websearch sites listed in the 'Useful websearch sites' section.

Tips for using mail order & home delivery services

Organic parents will find that shopping for organic food and products by phone, fax, internet and mail is convenient, time-saving and practical. The following are a few tips for making the most of organic mail order and home delivery services.

• **Shop around**: there is a variety of relevant organic products out there offered at varying prices. Check out websites and call the companies to compare quality, service, payment arrangements, products and prices (use the Products Guide). Many companies have annual sales, or send out discount vouchers with first orders, while others allow you to defer payment to help your budget—ask! Once they have your details, you'll be included in future mailouts of new products, services and catalogues.

• **If you're unsure about buying products** of which you'll need a large quantity, such as reusable nappies, firstly buy or hire a trial pack so you can try them out, or ask to talk to someone who uses them or sells a selection of them, to get their opinion and advice. Many companies will let you return unused items if they're not suitable for your requirements (return them immediately) and either refund or give you a credit against a new order.

• **Confirm the date you'll receive goods**, and where they'll be left if you're not home (ideally with a neighbour). Keep a copy of your order (including the date sent, any reference number and cheque number), and always note the person's name you dealt with for future reference, in case of any discrepancies or a lost order.

• **Ask the company to advise you if there's going to be any unexpected delay**. As some clothing and bedding orders can sometimes take several weeks to arrive, make sure you order early so they arrive in plenty of time for a new baby.

• **If you're not satisfied with products** when you receive them return them immediately and/or ask for a refund or credit against another order. If products perform poorly when you've followed all the correct washing/use instructions, inform the supplier, ask for a refund and send the product back straight away, with written reasons. If you experience similar problems again on the second order, return the goods and try another company. Tell the first supplier why you won't be placing another order with them. All organic goods should be quality goods.

• **Don't hesitate to ask for proof of organic certification**—by law companies must be able to provide this, even for imported goods (see A Brief Shopper's Guide).

• **Don't hesitate to ask suppliers whether they can provide you with products you want** if they don't have them listed—they may not stock such items simply because no one has previously asked for them.

• **Check on delivery charges**. Some companies charge for small orders while larger orders are delivered for free, and there may be a special charge if you want goods immediately. (For nationwide organic food home delivery, which can include ready-made babyfoods, charges may be included in the price if you order a vegetable box, but not if you just want to order other items—such as a large bag of carrots for juicing—on their own.)

• **You'll generally find that using a local or regional organic food delivery company is cheaper than a nationwide supplier.** However, a nationwide company can be more convenient as some operate seven days a week, deliver several times a week rather than only once or twice, and are more likely to provide babyfoods. Ask where the food comes from, how fresh it actually is, when it is harvested and packed, and the time delay from these stages to the time you receive it.

BABYFOODS

Breastfeeding for as long as possible, and making your own fresh and mainly raw or lightly steamed or lightly cooked organic babyfood, should always be first preference for your child's optimum nutrition (see Organic Food, Health & Nutrition Tips in Chapter Three). But breastfeeding doesn't always go smoothly, may not be able to be continued for various reasons, or perhaps cannot even be started. And parents are busy, often tired people with many commitments who travel around a lot, so it's not always possible, convenient or practical to prepare fresh food to achieve this.

The good news is that there are now many healthier alternatives to regular babyfoods (found to contain multiple pesticide residues), which parents can easily buy. There are at least four different brands of infant and follow-on formula milks, and at least ten brands of ready-made organic babyfoods available as a back-up to your own healthy meals. These require little or no preparation, and can be easily carried with you in a bag or push-chair when out and about. Meals in jars or frozen meals can be stored for longer; packet foods such as cereals can be resealed or folded up for future use; and jar foods, once opened, can be refrigerated for between 24 and 48 hours. Fresh, ready-made babyfoods can be refrigerated for up to fourteen days, and even frozen. These convenient organic meals for your baby can now be found in healthfood and fine food shop chiller cabinets, and on shelves in Boots, major supermarkets and healthfood stores. Many are also available through organic food home delivery schemes, or by mail order direct from the manufacturer.

Vegans, vegetarians and meat eaters are catered for, although unfortunately no organic company is currently offering alternative vegan protein such as tofu or seaweed in baby meals for older babies. For meat eaters, nutritionists advise not to introduce meat dishes until seven to eight months, and then sparingly, preferring organic fish and poultry to red or cured meats such as ham and bacon (not recommended for babies). Ideally, choose whole grain cereals rather than refined white baby rice and cereals (use from eight months on), but again don't over-use them, as they are harder to digest than fruit and vegetables.

Organic babyfoods are certified to be grown without pesticides or artificial fertilizers, are GM-free, are minimally processed, and contain higher quality ingredients than regular brands and are also free of added salt, sugar (with a few exceptions), and the many artificial and unnecessary additives found in many conventional babyfood brands. These include fillers, emulsifiers, thickeners (such as vegetable oils and chemically modified cornflour), flavourings, colours and added multiple sugars (three or more can be found in some regular meals), most of which have little or no health benefits and can in fact be detrimental to health. Some conventional brands use vegetable oils and maize ingredients which may come from GM crops, and vegetarian cheese, which contains rennet from a genetically engineered source, rather than calf rennet. Organic vegetarian cheese used in organic babyfoods (and in fact any organic food) uses an alternative GM-free and animal-free ingredient.

However, always read the label on organic brands. While most babyfood meals are sugar-free, some do contain sugar (although in much lesser amounts than regular brands) or honey, and some contain baking powder. Maltodextrin (a maize starch/flour), commonly used as a thickener and sweetener in regular babyfoods, is unfortunately also used in organic formula and follow-on milks—it is low in nutrients and becomes a gluey paste with water.

With organic babyfood meals, you'll find that the ingredients after which the product is named make up all or nearly all of the product (with leading brand Baby Organix labelling the percentage of all ingredients). With regular babyfood brands, this is often not the case. With Heinz Organic range, for example, the main ingredients make up a very variable percentage of the whole. Their Organic Apple Dessert jar from four months contains 99% apple; but their Vegetable Mash with Lamb contains skimmed milk (percentage unspecified), 28% vegetables (potatoes, parsnips and onions), water (percentage unspecified), and only 8% lamb. In their new Organic Simply range, Organic Pear and Pineapple is 60% pear and 40% pineapple. Organic Pumpkin and Sweet Potato contains 50% pumpkin, 25% potato and 25% water. Details of their product range is on the Heinz Tiny Tums website at www.tinytums.co.uk, including the nutritional content of all products.

As from June 2002, all Sainsbury's own label babyfood (a total of 185 products) is organic. Their babyfoods now include percentages of ingredients (having changed their policy since 1999, when the first edition of this book was published). This reveals that their organic Vegetable & Chicken Risotto contains only 20% vegetables (carrots and tomatoes) and 8% chicken. However, their Organic Baby Fruit Purée Peach contains 60% Peach Purée, 36% Apple Juice, plus Rice Flour, Rice Starch and Vitamin C.

Many of these brands also enliven their labels with 'enriched with vitamins and minerals' and add up to thirteen synthetic vitamins and minerals to such meals (a food industry practice used on many children's foods, which the Food Commission calls creating 'fortified junk'; it tricks parents into believing they're giving their children healthy food). Organic products either don't use these at all (after all, they're still in the organic food), or add one or two key ones.

Organic babyfood now accounts for almost half of the market in the UK. HIPP remains the market leader with around 20% of organic brand sales—more than twice any other organic brand. Second is Heinz Organic, followed by Baby Organix, Cow & Gate Organic, Sainsbury Own Label Organic, and Boots Own Label Organic. A diverse range of organic babyfood products is now available, including puréed wet simple first foods, breadsticks and rusks for teething babies, dry packet cereals, and many wet savoury meals with mixed vegetables, pastas and meats, fruit puddings and fruit and yoghurt desserts. Ask your local healthfood store or organic home delivery service to order in products if they don't stock them, or order direct through a mail order service.

Most organic babyfood companies can provide you with weaning guides and charts, nutritional information on their products, and useful recipes for making your own babyfood. Some even have a nutritionist you can talk to about products and any problems or queries regarding weaning. Try to reuse the glass babyfood jars as much as possible for your own mixtures, or at least recycle them to reduce packaging waste.

Most leading organic babyfood companies (in line with their conventional cousins) package their baby rice and cereals in aluminium foil, and small rice cakes also usually come in foil packs. Top nutritionists now advise people concerned about their health to avoid aluminium foil in food packaging, as they claim that the aluminium can pass into the food. Raise this issue with companies and favour products packaged in natural cellophane or paper.

Formula milks

Organic breastmilk is nature's food for your baby. It is recommended as the best (and only) food at least until weaning at between four and six months, and ideally should be used as part of your baby's diet for one year or more (see Breastfeeding in Chapter Two). However, not every mother is physically able to breastfeed her baby, and some want to be able to mix feed (some breast and some bottle), especially if they work or have other commitments. Mothers found to be HIV-positive or with other serious diseases or health problems may be recommended not to breastfeed. Thankfully there are now several organic formula milks available for these mothers and babies in powder form in healthfood shops, or available via mail order. However, at present there are no organic vegan formula or follow-on milks for vegan mothers and babies. Organic ingredients are listed in order of quantity (largest first); added vitamins and minerals are not organic.

Babynat* Infant Milk Made in France by Vitagermine, this organic infant formula is suitable from birth to six months. It is based on organic cow's milk and contains demineralized whey powder, organic cold-pressed vegetable oils (palm, rapeseed, olive and sunflower), skimmed milk, maltodextrin from maize, vanilla flavouring, soya lecithin and minerals and vitamins; it comes in a 400g tin for £5.45. Available direct from Organico Realfoods Ltd* and many healthfood stores.

Eco-Lac* Formula Milk Made in Holland, this milk is suitable from birth to six months. It is based on cow's milk with organic vegetable oils, and comes in small boxes for mothers looking to supplement breastfeeding or wanting to stir into baby cereals. It contains whole milk powder, whey powder, cream, maltodextrin, vegetable oils and added vitamins and minerals. For a 450g box, the milk retails for £5. Available in independent healthfood stores, contact importers DeRit for stockist information. Phone 01422 885523.

Holle* Instant Baby Milk Made in Germany, this formula milk is based on organic cows' milk and is biodynamic. It contains full fat milk, whey powder, cream, organic vegetable oils for fatty acids and maltodextrin (maize starch/flour), added vitamins and minerals. A 400g tin costs £4.95 and is available mainly by mail order and in a few healthfood shops. Available by mail order from Elysia*.

Nanny's Goat Milk Infant Nutrition Although, not organic, Nanny's goat milk formula provides a good alternative for babies who can't digest cow's milk (goat milk is more similar to human breastmilk than cow's). It comes in 400g and 900g resealable tins and is also suitable as a cow's milk substitute for children and adults. Sold in many independent healthfood stores. Call Nanny's helpline 0800 328 5826, for information and stockists.

Follow-on milks

While breast is always recommended as best for baby's health (see Breastfeeding in Chapter Two) these follow-on milks are available as an alternative or supplement, and come in powder form. Suitable for babies from four months of age, they can be used in place of breastmilk with weaning cereals. You'll find them in healthfood shops (and some brands are now in major supermarkets), and can get them by mail order.

Babynat* Follow-on Milk Made in France, this organic milk is suitable from six months. Based on organic cows' milk, it comes in a 400g tin for £4.45. It contains maltodextrin from corn, skimmed milk, demineralized whey powder, cold-pressed vegetable oils (palm, rapeseed, olive, sunflower), natural vanilla flavouring, soya lecithin and added minerals and vitamins. Available from Organico Realfoods Ltd.* by mail order and many shops.

Eco-Lac* Follow-on Milk Made in Holland, this organic cow's milk drink comes in a 400g box for £5 and is suitable from six months of age. It contains skimmed milk powder, maltodextrin, vegetable oils, starch, and added vitamins and minerals. Available in independent healthfood stores, contact DeRit for stockist information. Phone 01422 885523.

HIPP Organic Follow-on Milk Drink Made in Germany, this organic cow's milk-based drink comes in a 900g box for £6.99 from four months. It is part of the extensive market-leading HIPP range of organic babyfoods and is available from most supermarkets. It contains organic skimmed milk, organic sweet whey powder, organic starch, organic vegetable oils and added vitamins and minerals. Contact HIPP Nutrition UK for further information and stockists. Phone 0845 0501351 www.hipp.co.uk.

Weaning & first-year foods

During and after weaning, several breastfeeds (or bottle-feeds) a day are still recommended until at least one year of age. It is also advisable to offer your baby pure water (boiled, bottled or preferably filtered) between meals, and you can also try watered-down raw organic vegetable (such as carrot) or fruit juices (one a day—see Organic Food, Health & Nutrition Tips), or the odd organic processed juice watered down. And don't be in a rush to wean your baby: make it a gradual, unforced process, to allow his developing digestive system time to adapt.

Although they are very convenient, you should limit the number of cooked, processed commercial babyfood meals you give your baby as they will lack the enzymes and many of the other quality nutrients found in raw, fresh foods. Even though organic babyfoods are highly preferable to regular brands, remember that they are still processed foods made in factories. (Jar meals are heat-sterilized, and there have been concerns about lid plasticisers leaking into some organic and regular brand jar meals. Although there are strict regulations on permissible amounts of plasticisers in foods, in November 2002, an EU regulation was passed giving lid manufacturers three years to come up with totally safe alternatives. Some companies, including Baby Organix, cook their babyfood in a way that doesn't allow contact between the heated lid and the food inside.) They should never replace your own fresh home-prepared foods, but instead be used as a convenient back-up or a handy supplement (such as cereals). Try several different brands and recipes to find which you and your baby prefer. Introduce simple one fruit, one vegetable, one pulse or one cereal meals first (from four months), then gradually move on to meals with two ingredients, and then on to more adventurous meals (say from six to seven months) with many different wholefood ingredients. This lets you find out what your baby prefers, and also pinpoint any allergens, so when using commercial foods, stick with simple ones first. Wholegrain foods, such as rice, pasta and noodles, can be given to babies from eight months old, as until then their digestive systems are unable to cope with such high-fibre foods.

You can increase the nutritional content of jar meals by blending with fresh produce (if you have some, but not enough to make a full meal for your baby), a small amount of finely milled organic seaweed (such as Seagreens or Clearspring Sea Vegetables), essential fatty acid oils (such as Udo's Choice or Essential Balance), or fresh carrot juice, vegetable juice, or even a small amount of wheatgrass juice or powder (such as Sweet Wheat or Four Seasons). You can also add such oils and other foods to cereal meals, and use rice, almond, soya, oat and pea milks and coconut milk or cream in small amounts, either on their own, or blended for older babies to give different tastes and nutrients. If you drink cow's milk, consider buying organic 'green top' unpasteurized milk from a reputable organic farm—although some nutritionists advise avoiding cow's milk for at least the first year as it is a potential allergen (this could be due to the processing, so give green top milk a try).

And get your baby used to taking ambient food (i.e. at room temperature) as soon as you can—heating and cooking destroys nutrients and gives you less flexibility for preparation. Be aware that jar meals containing cocoa and chocolate (by Cow & Gate and Bio Bambini) will contain the stimulants caffeine and theobromine, which are not recommended for babies. See also Organic Food, Health & Nutrition Tips in Part One.

Baby Organix Launched in 1992, this UK brand is the country's second largest selling organic babyfood range, with over eighty baby food products, and is increasing the number of its lines all the time. Founder Lizzie Vann won the Soil Association 2002 Organic Trophy for her efforts to promote safe and healthy food for children, and the company won the *Here's Health* Magazine 2002 Junior Wellbeing award. It claims it was the first UK company to fully label quantities of ingredients—it was certainly the first babyfood company to do so.

Baby Organix uses refined (white) flours and rice, and no sugar (except malt in its breadsticks) or thickeners. It clearly displays the 'No GM ingredients' message on its jar lids. The company was founded by Lizzie Vann and Jane Dick to provide the first healthy processed babyfood alternative to the conventional brands. Lizzie (a stepmother of two) has been a vocal campaigner for healthier food for babies, and for full ingredient labelling.

For older babies, Baby Organix has thirteen cereal recipes in 120g packets for £1.95, ranges of dry pasta shapes, a large selection of fruit and savoury jar meals (190g for 85p), dry packet muesli, porridge and rice meals, and breadsticks. Baby Organix meals include Mild Vegetable & Coconut Korma, Vegetables & Pork with Apple Sauce, Fruit Compote, Banana Porridge and Tomato & Chicken Casserole. Packet meals include Baby Rice with Garden Vegetables, Apple & Raspberry and Banana & Orange Yoghurt. Pasta packs contain Alphabets, Ducks or Stars. Organic Olive Oil Breadsticks for teething babies contain white organic wheat flour, organic extra virgin olive oil, yeast and sea salt, and are said to be ideal for babies from six months learning to hold foods and chew. The breadsticks cost £1.14 p for a 100g box.

All Baby Organix products have a shelf life of two years. Call their Nutrition Line for product information including stockists, mail order (for dry goods only), a colour guide to weaning, and useful recipes. Baby Organix products are extensively available in healthfood shops, Boots, independent pharmacies, Mothercare World and all major supermarkets. Much of the range can be ordered via mail order from The Organic Shop*, Simply Organic* and Goodness Foods*. Phone 0800 393 511 www.babyorganix.co.uk lizzie.vann@organixbrands.com.

Babylicious Recipes are produced from organic ingredients gently cooked in small batches before puréeing. The purées are then quick frozen into small portions to maintain the flavours and nutrients and provide the ultimate in convenience for tasty, quick meals for your baby, with no waste because you simply use what your baby will eat. Their Pure and Simple Organic Vegetable Purées (from 4 months) include cauliflower & broccoli and carrot & potato; Pure and Simple Organic Fruit Purées (from 4 months) include apple & pear with cinnamon, and banana dessert; and their Exciting Organic Vegetable Purées (from 5/6 months) include a root vegetable medley and cauliflower cheese. Each pack has two different recipes, price £3.79 for 500g. Available from Sainsburys and some independent stores. Phone 0208 998 4189 www.babylicious.co.uk info@babylicious.co.uk.

*Babynat** Imported from France, Babynat has a vegetarian range of seventeen organic babyfood products including simple fruit and vegetable recipes in 130g jars suitable from 4–6 months for 89p. Babynat also supplies a nutritional guide with its products. Meals include Organic Peaches, Organic Apple & Blueberry, Organic Courgettes & Pumpkin and Organic Provençale Vegetables. Babynat Teething Biscuits (1.95) are suitable from five months and contain wheat flour, cane sugar, egg, corn starch, baking powder (ammonium bicarbonate) and sweet orange essential oil. The jar meals contain no added sugar. Available from shops, direct from distributor Organico Realfoods and other mail order companies. Phone 0118 951 0518 (Organico Realfoods) www.organico.co.uk info@organico.co.uk.

Bio Bambini A range of eighteen biodynamic babyfood jar meals from the German company Sunval. Available in three stages for easy weaning, there are simple recipes from four months onwards, including Parsnip & Potato and Baby Fruit Cocktail; meals introducing meat and cheese from six months including Beef, Carrot & Potato and Apple & Blueberry Pudding; and meals from eight months including Turkey with Rice & Sweetcorn, Semolina with Vanilla and Choco Banana Pudding. Shelf life of unopened meals is 36 months; agave or pear juice concentrate, maple leaf syrup and grape juice are used as sweeteners in some fruit and pudding meals. Some meals contain sunflower oil and herbs. The jar meals, which all contain water, range in size from 125g to 190g and in price from 69p to £1.40. They are available in health food stores, local pharmacies and mail order from Eco-babes* and Spirit of Nature*.

*Boots** *Organic* Boots now has organic babyfood products available in stores and through Boots online mail order catalogue. Jar meals, some of which contain wholegrain rice starch and water, are available in 125g for 55p and

190g for 65p. Meals include Mixed Fruit Purée, Vegetable & Lamb Casserole, Apple & Blueberry, Spaghetti Bolognaise, Apple Crumble, Vegetables & Steak, Creamed Porridge. There are also cereals including Organic Fruit Porridge and Organic Fruit & Muesli in 250g cartons for £2.45, and Organic Baby Rice in 180g cartons for £2.69. Dried meals in 180g boxes for £2.25 include Potato & Leek, Sweet Corn & Vegetables. Phone 0845 070 8090 www.wellbeing.com.

Eco-Baby This small organic range from Dutch company Nature's Home is imported by DeRit. Eco-Baby has five wholegrain baby cereals, made in Holland: Rice meal (from four months) comes in 200g boxes for £2, Muesli (from seven months) in 200g boxes for £2.36 and Mixed cereal (from seven months) for £2.00, Gluten-free in 200g for £2.36 and Banana in 200g for £2.26. Muesli contains ground maize, barley, rice, apple, maize malt, hazelnuts and almonds, and Mixed Cereal contains ground rice, maize and barley. All Eco-Baby products are wheat-free, as some nutritionists say that most children under one year old will benefit from a wheat-free diet: it's difficult to digest, and also a known allergen. Grains for these products are milled in a restored 19th century windmill and processed in a small modern factory which has been painted with organic paints. The babyfood is pre-cooked, using a method similar to pressure cooking at home, which apparently opens up the grains and makes them easily digestible. Packaging is in biodegradable cellophane, recyclable card which is chlorine-free, and uses soya-based inks. Eco-baby is available from some shops and mail-order companies. Call DeRit for stockists. Phone 01422 884629 (DeRit) www.natureshome.nl (Netherlands site).

Familia Made in Switzerland, this organic Swiss Muesli babyfood (there's also an adult version) with no added sugar comes in 340g packets for £1.80 and can be found in independent healthfood stores.

Feed Me, Feed My Baby Independent family-owned catering business using only organic ingredients. They will create tailor made menus and come to your home to cook up to one month's worth of wholesome organic food. The service is specifically designed for new parents who have their hands full with their new arrival, but want to continue to eat healthful, organic meals. The service costs £27.50 per hour for a minimum of six hours and a £50 deposit is required to register for the service. Phone 020 7511 1627 Mobile 07930 327263 www.feedmefeedmybaby.co.uk.

HIPP Made in Germany by the Hipp family for many years, HIPP is now the leading organic babyfood brand in the UK, offering babyfood products with organic fruit, vegetable, meat and dairy. The range includes dried breakfast cereals (using refined [white] rice), vegetable or fruit-only jar meals, meat (chicken, pork, turkey and beef), dairy, pasta and vegetable meals, low-sugar rusks, yoghurt and fruit desserts, and bottled fruit juices. Foods are labelled from four months, from seven months, and toddler meals from twelve months. Some breakfast cereal meals contain organic sugar and/or honey, and rice starch is used in many jar meals. Apple & Grape and Redberry fruit juices (juice with mineral water) cost 75p for 200ml; babyfood jars of fruit, vegetable, meat and pasta recipes and desserts come in three sizes for 59p (125g), 69p (190g) and 99p (250g); and refined cereals come in 150g bags for £1.89. Meals include Spring Carrot & Sweetcorn, Vegetable & Chicken Risotto, Spinach with Potatoes & Cheese, Fruit Duet Mango & Banana with Yogurt, and Creamed Porridge Breakfast. Low Sugar Rusks are 99p for 100g. Toddlers' meals (in wide-necked jars) include Chicken Nuggets with Rice & Vegetables and Mini Beefburgers with Tomatoes & Rice. HIPP babyfoods are available in independent health food stores, most major supermarkets and many mail order companies. Contact HIPP Nutrition UK for further information. Phone 0845 0501351 www.hipp.co.uk.

*Holle** Made in Germany by a Swiss company for over sixty years, Holle has a range of biodynamic (Demeter certified) wholegrain baby cereals suitable from four months in 250g cartons costing £2.95, including Baby Rice, Baby Millet, Mixed Cereal Porridge, Oat Porridge, Semolina, Spelt Porridge. Holle also sells baby butter rusks in 150g packs for £3.25 (they can be crumbled into purées or used during teething and beyond); they contain bran-reduced wheat flour, whole wheat flour, honey, butter, yeast and sea salt. Holle products are only available in a few healthfood stores, but can be obtained direct from Elysia by mail order. All cereals come in recycled carton packaging and the shelf-life is one year. Phone 01386 792 622 (Elysia mail order) www.drhauschka.co.uk enquiries@drhauschka.co.uk.

Johanus Imported from Germany, these biodynamic babyfoods come in nine flavours, in 190g jars for 79p. They are available from leading healthfood stores including Olivers in Kew and Infinity Foods in Brighton. Meals include Apple, Oat & Blueberry, Carrot & Pea, Pear, Apple, Oats & Honey and Spinach & Rice. If your store doesn't stock them, ask them to get hold of them via Infinity Wholesalers of Brighton or Marigold in London. Available by mail order from Cook's Delight (seeing Leading Stores).

Kallo Kallo makes Organic Rusks with a rabbit design for babies from four months, price £1.49 (containing wheat flour, 7% raw cane sugar, sunflower and palm oil and fats, soya flour, sodium bicarbonate, ammonium bicarbonate and thiamin—vitamin B1); Organic Biscuits with jungle animal designs and inside colouring card and animal facts for children from ten months for £1.49 (ingredients as above with additional pasteurized whole eggs, honey and skimmed milk powder); and plain Organic Junior Rice Cakes for 59p. Made in Belgium, they are widely available in independent healthfood stores and supermarkets. Kallo also have adult-sized organic rice cakes suitable for older children, including banana, caramel and marmite flavours. www.kallofoods.com.

The Organic Baby Food Company 100% organic made-to-order frozen baby food delivered to your door. Order what you want from a list of options depending on the age of your baby. For 3–5 months: vegetable purées (swede, parsnip, green bean) and fruit purées (apple, pear). For 5–7 months: vegetable purées (leek, tomato, sweet red pepper, beetroot) and fruit purées (apricot, grape, plum). 8 months and over: pasta sauces, vegetable purées (potato and broccoli, carrot, parsnip and swede) and puddings (mango, apple and cinnamon, melon and peach). All food is frozen into cubes within two hours of cooking. A bag of 14 cubes costs £3.50, 28 cubes costs £7. Phone 01666 505616 Mobile 07773 404744 www.tetbury.com/baby.

Truuuly Scrumptious Organic Baby Food A winner in the Soil Association's Organic Food Awards 2002, this company is run by mothers Janice and Topsy. They make and sell fresh, frozen baby food with no additives, salt, sugar or fillers in resealable tubs. Small pots (100g) are suitable for 4 months and older and include butternut squash, swede, carrot or broccoli and cost £1.10. Medium pots (160g) for 6 months and up include sweetcorn chowder, vegetable gratin and salmon and broccoli pie and cost between £1.60 and £1.75. Large pots (220g) for 12 months and up include vegetable and lentil bake, tomato and herb pasta and winter vegetables and chicken and cost between £2.10 and £2.30. Sweet meals of single or mixed fruits are available in each size also. Truuuly Scrumptious can be found in several health food shops (list on website) or ordered direct from the company for delivery in Bath and London. Also available from Graig Farm. Phone 01761 239300 www.bathorganicbabyfood.co.uk sales@bathorganicbabyfood.co.uk.

* * *

Conventional baby food manufacturers Cow & Gate and Heinz both have organic ranges of baby food, available through supermarkets including Waitrose, Sainsburys and Tescos, as well as Boots. There are also own brand ranges from Sainsburys (now 100% organic) and Boots.

BABY GEAR

The natural cotton and woollen fibres used in organic clothing are the safest, softest and most luxurious for your baby. They have not been grown or treated with pesticides and artificial fertilizers, fumigated or genetically engineered. The fabrics are not treated with flame-retardant chemicals (which have been linked to cot death), organophosphates or other pesticides, formaldehyde, strong chemical dyes (which can contain heavy metals), bleaches (which can leave behind dioxins) or stiffeners during production, and do not contain residues of these substances which could harm your baby or give him allergies. Furthermore these natural fibres allow the skin to breathe and naturally absorb moisture, unlike synthetic fabrics, which are derived from petroleum and do not allow for air and moisture circulation.

By buying organic cotton clothing for your baby you are not contributing to conventional cotton-growing, which accounts for about a quarter of the world's pesticide sales—it is one of the most heavily sprayed agricultural crops, and contributes to poor health in farm workers and their families in developing countries (where chemicals which are banned for use in the UK due to their toxicity to humans and wildlife, are often used). Genetically engineered cotton poses a further grave risk to the environment from the potential transfer of new genetic material by pollen, the creation of superweeds and superbugs, potentially harming or killing insects and birds, and increasing pesticide usage.

Organic wool has not been dipped in organophosphates or sprayed with other treatments or chemicals while on the sheep's back or during processing, and so has not endangered human health. Organic sheepskins have not been tanned using harmful artificial chemicals, and organic sheep are raised with higher animal welfare standards than in conventional farming. Organic clothing is not made with child labour, nor in sweatshops with poorly paid labour.

The people behind organic companies offering organic babywear are generally families who have sought alternatives for their own babies and then starting making or selling the tried and tested products they liked, and that their children wear, to help other families. If you can only afford a few items, choose ones which will be closest to your baby's skin, such as nappies, short body suits and vests. Wash them with environmentally friendly cleaning agents (featured on pages 156–8).

The Soil Association has recently brought out (in 2003) new standards for organic textiles. To be certified, clothing (both domestically produced and imported) must contain a minimum of 70% organic materials derived from plant or animal sources, with an ideal of 100%. Sheep will have to have been reared organically for 12 months prior to having their wool certified.

Nappies

Many brands of reusable unbleached cotton and wool organic nappies are now available by mail order; although there is an initial cost outlay, they work out much cheaper in the long run than using polluting disposables. Organic resusables are safer for a baby's delicate skin (especially if your baby already suffers from eczema) and developing internal organs. Even nappy rash can be reduced by wearing cloth nappies, which allow the skin to breathe, and are soft against a baby's delicate skin. Organic nappies are free from gels, glues, chlorine bleach, plastics, formaldehyde, dyes, artificial perfumes, chemical lotions and other toxins found in plastic and paper disposable nappies, or used in their production. Toxic organochlorines, such as cancer-causing dioxins, can be created in the production of disposables in both wood pulp and plastics manufacture. Chemical additives are used in plastics, and in the past one of these turned out to be an oestrogen mimic associated with sex changes in fish. This was a wetting agent used on the inner netting layer to stop urine staying on the surface—it has now been withdrawn. Events like this remind us that disposables and other plastic items are industrialised products which could have unexpected effects on the environment and even on health.

A study in 2000 suggested that decreasing male fertility may be due, in part, to babies' testicles overheating in plastic nappies. (Testicles must stay a few degrees cooler than body temperatures in order to develop properly.) Recent research in the United States has shown disposables to emit a number of noxious airborne chemicals as they outgas (e.g. as soon as they are removed from the bag), including toluene and benzene, both of which are linked to cancer. In tests with laboratory mice, these emissions caused a range of respiratory symptoms, which led the researchers to conclude that fumes from disposables could be linked to asthma— a growing health problem in children in the UK. Small blobs of the chemical absorbent gel used in disposables can often be found in the baby's groin area, and there's a risk that babies could ingest the gel by putting the blobs in their mouth.

Organic nappies are as beneficial for the environment, wildlife and waterways as organic food—not only because of their benign production methods, and the fact that they use far less energy to make, but also because using them helps to reduce the huge number of disposable plastic and paper nappies estimated by environmental groups to be dumped in the UK (about 4% of household rubbish alone, with many of them containing live vaccines), creating over 800,000 tonnes of rubbish costing an estimated £100,000–200,000 a year per average local authority to collect and dispose of in rapidly filling landfills. Then there are the estimated more than seven million trees that are cut down to make them each year (planted instead of native forest, displacing native plants and animals, and creating factory waste and pollution in processing). Plastics, which are manufactured from oil (about one cupful for each disposable nappy) take hundreds of years to degrade (if at all) and the parts which do degrade produce methane, a powerful agent in global warming. According to the Real Nappy Association, by using reusable nappies a family can save up to £600 a baby— 'real' nappies costing around £400 for the first baby including purchase, washing agents, electricity, and wear and tear on your washing machine etc, compared with about £1000 for disposables (savings can be even greater with some organic nappy systems and clever use of washing agents). Reusable nappies are promoted by the Real Nappy Association and Women's Environment Network all year round and especially during Real Nappy Week; many local councils now subsidize the cost of real nappies for new mothers, as they realize it is an ideal way to avoid the expensive landfill tax. If you're unable to cope with the washing load at any stage, for those who can afford it there are nappy-washing services available in most areas.

At least twenty UK companies (many of them small family concerns) currently offer ranges of European, US and UK-made shaped, pre-fold or tie-up organic nappies in unbleached cotton and organic or chemical-free wool. There is a bewildering variety of nappy styles so it's a good idea to get a sense of what works best for your baby's needs before buying a whole lot. For valuable advice, contact or see the websites for The Real Nappy Association, The Nappy Lady, Cuddlebabes* and Eco-babes*. Each of these companies provide information on the different styles of nappies. The Nappy Lady gives reviews and anecdotes from parents on her website and Eco-babes has trial nappy kits for hire.

*Beaming Baby** Offers organic Imse Vimse flannelette shaped nappies, £4.50 each. Also in the Imse Vimse organic range from this company are cotton towelling shaped nappies (£8.49), Snug-to-Fit one size with Velcro (£8.99), and towelling popper fit (£9.49). Also offer non-organic nappies and covers including Motherease rikki covers with Velcro for £7.69.

*Born** Sells Gossypium-designed organic nappies in one size—adjustable with poppers—for £8.95. Also Disana tie nappy from £1.75 and muslin squares, £1.75 for one and £5.25 for a 3-pack. Felted wool cover with velcro cost £15.00, woven wool ones are £17.00. Born also sells non-organic nappies and covers and a range of nappy accessories including liners, buckets, bags, Nappy Fresh, Aquaball and Tushies disposables.

*Clothworks** This Bradford-upon-Avon company makes organic terry squares for £6 each with brushed cotton pads for £2.50.

*Cotton Comfort** This company which makes hypoallergenic clothing for adults and children with irritable skin conditions sells an organic fitted nappy, which costs from £9.50 for 2 nappies.

*Cuddlebabes** Offers both organic and non-organic ranges of nappies. In the organic range: Under-the-Nile shaped one size with snaps for £9.95 (£9 each if purchasing four or more), Disana tie nappy for £1.50 each, flannelette liner for £1.50, Muslins (10 pack) for £13.50, Snug-to-Fit one size for £8.95. Natural wool covers cost from £8.95, fleece covers from £6.99 and waterproof synthetic covers from £5.99. Nappy accessories available including washing nets, buckets and washable wipes.

*Eco-Babes** Eco-Babes has Popolino organic cotton shaped Ultra-fit terry nappies in three sizes costing from £9.50. Snug-to-Fit offers the same design in felted cotton for £8.99. Under-the-Nile one size terry nappy costs £9.99. In their organic Imse Vimse range, the organic terry wraparound (to fold into cover) costs £8 and a four pack costs £32. There are also knitted felted woollen wraps from £8.99, microfibre overpants in sizes small to extra large from £17. Eco-babes has a range of other nappy accessories including paper and silk liners, biodegradable nappy bags, nappy nets and washable wipes.

Eco-babes Kits and Agents Department This department and website is specifically for hiring nappy kits and finding real nappy agents in your area. A great way to see what nappy systems work best for you and your baby. Kits consist of used, sterilised nappies and can only be hired, not purchased. Newborn and Trial kits contain some organic nappies. Also available is a Wraps kit and Trainer Pant kit. They provide a list of real nappy agents, which you call to hire the kits. Kits cost from £6.25 per week with a £60 refundable deposit. P&p is £5.95 and the kits must be hired for a minimum of 2–4 weeks (depending on kit). Phone 01833 640 400 www.ecobabes.freeserve.co.uk kits@ecobabes.com.

Ethos Amidst a range of non-organic nappies and nappy accessories, Ethos sells one organic nappy: Snug-to-Fit with poppers one size for £8.95. They also sell Nappy Fresh, Aroma Kids products and Tushies wipes. Phone 0845 0900 184 www.ethosbaby.com enquiries@ethosbaby.com.

*Green Baby** This mail order company and shop in Islington sells five brands of organic nappies including Imse Vimse, Green Baby and Under the Nile starting at £8.95. All nappies are also available in trial packs. They also sell non-organic nappies, covers and accessories like nappy bags (throw dirty nappy in and wash the whole bag in the washing machine).

*Greenfibres** Greenfibres offers German-made Disana organic cotton knitted tie-up nappies in sizes from newborn to three years for £ 20.30 for a pack of ten. Organic and biodynamic woollen overpants cost £11.10 and come in four sizes, and organic cotton inserts cost £17.25 for a 10-pack. There are also raw silk liners costing £9.10 for a pack of 3. A trial pack with paper liners costs £4 without woollen overpants, or £14 with one. Greenfibres also supplies a waterproof sheet made of natural rubber encased in untreated cotton, to change your baby on instead of a plastic change mat, for £18.20 (£33.90 for a large one). Also an organic cotton changing mat filled with organic cotton wadding. Merlin changing mat (matches Merlin bedding—see Greenfibres in Bedding section) for £49.50 (75x85cm).

Little Earthlings This company has a range of conventional cloth nappies as well as the organic Imse Vimse shaped terry nappy for £8.50 and the organic Snug-to-Fit (with poppers and Velcro) for £8.99. Order online or by phone. P&p is free on orders over £100. £5 gift vouchers also available. Phone 028 2954 1214 www.littleearthlings.com info@littleearthlings.com.

Lollipop This baby mail order company sells several types of organic nappies including Disana tie-up (£1.50), Imse Vimse shaped (£8.50), Under-the-Nile one size (£9.99) and Snug-to-Fit (£8.95). Phone 01736 799512 www.teamlollipop.co.uk enquiries@teamlollipop.co.uk.

Nappi Nippas Simple fastening system for cloth nappies requiring no pins. A single Y-shaped fastener costs £1.79 and pack of 3 costs £3.99. These are very popular and can be purchased through many mail order companies including Eco-babes*, Green Baby*, Born*, The Nappy Lady, The Nice Nappy Company* and Twinkle, Twinkle*. Phone 01736 351263.

The Nappy Lady Morag Gaherty sells three types of organic nappies: Poplino multifit (7 to 35lbs) for £7.99, Under-the-Nile one size for £9.99 and Disana tie-up for £1.50. She sells a variety of non-organic nappies and accessories including Disana wool wraps costing from £11.50 and Bumby wool covers for £17. Order by post, email or phone. Delivery is usually 1–2 days. P&p cost £1.99 on orders up to £30, £2.99 on orders £30–£50 and £4.99 on orders £50–£250. P&p is free on orders over £250. Phone 01622 739034 www.thenappylady.co.uk.

Natural Child* Sells two types of organic nappy: Green Baby Snug-to-Fit with velcro and poppers for £9 and Under-the-Nile one size with poppers for £10. They also offer non-organic nappies, Tushies disposables, liners, a nappy soak and disposable wipes.

Natural Collection* Offers a velcro fastening organic cotton reusable nappy, made in Germany. Prices start from £7 for a small nappy, and you can buy twelve from £75.50. Organic cotton nappy inserts increase absorbency—they cost from £3.50 for one or from £15.75 for five. A waterproof overpant made of microfibre (from £8.50) makes the system waterproof but breathable, and biodegradable paper liners are also available. A nappy trial pack costs £15 with one of everything.

For further details about companies marked with an asterisk (*) including how to order, postage & packing charges etc, see the Leading Mail Order Suppliers section on pages 164–167.

Naturally Nappies Carries two types of Disana organic nappy: Tie-up (3 for £4.50) and Shaped (from £5.25 each). Also Disana flat liners/hush cloths (3 for £4.50) and knit woollen wraps (£11). Naturally nappies also offers hemp and unbleached cotton prefolds (not organic) for £16.50 for a pack of four. The company also sells several types of non-organic nappies and disposables, wraps and accessories. Order by phone or email. Phone 0870 745 6641 www.naturallynappiesuk.co.uk. Order online at naturallynappies.com.

The Nice Nappy Company* This company sells a wide range of non-organic and organic nappies and accessories. Organic nappies include: Disana tie-up, 3 for £4.75 and Imse Vimse shaped terry (one size), 1 for £8.50 or 4 for £32. Wool pull-over cover costs for £8.99.

Schmidt Natural Clothing* Schmidt is one of the longest running organic family clothing companies (thirteen years), and is run by father Glenn Kooitzki Metzner in East Sussex. He claims you can spend just £150 to £200 from birth to toilet training with its wraparound nappy system—about a six month bill for disposables. Imported from Europe, Schmidt organic Disana knitted nappies consist of organic muslin or brushed cotton liners for extra absorbency, knitted organic cotton wrap-around nappies which tie-up at the front (one size fits all) and hand or machine knitted organic woollen pants (available in five sizes) or chemical-free felted wool outers. Trial packs with machine-knitted outer pants are available from £13.10, and nappies and their various parts are all available separately. A pack of three organic cotton knitted wraparound nappies costs £5.70. A pack of three organic cotton muslin squares costs £5.95, and brushed organic cotton liners cost £4.90 for a pack of three. Organic machine knitted wool pants for newborn to 6 months cost £9.45 each, hand-knitted pants cost £11.90, and felted woollen outers cost £14.90 (suitable up to nine months). Schmidt offers a two month payment plan for orders over £30. Paper liners are also available.

Spirit of Nature* Spirit offers two types of imported nappy systems. Organic shaped, brushed cotton nappies come with practical velcro fasteners. A muslin or hush cloth can be used for extra protection, and microfibre or woollen overpants (felted or knitted) make the system waterproof. Shaped nappies cost £5.98 each. Spirit also offers an organic cotton knitted tie-up nappy, which can be used with silk or cotton liners, for £4.79 for a pack of three (one size fits all). Spirit also has milled unbleached virgin wool overpants (£8.19 each), felted woollen overpants (£13.29 each), waterproof microfibre overpants (£6.95 each), muslin and organic cotton hush cloths (£3.29 for a pack of three), or organic twilled cotton squares (£4.29 for a pack of three), and paper liners. A trial pack of the tie-up nappy system costs £12.59 (two nappies, two hush cloths, four paper liners and one woollen overpant), and a trial pack of the velcro nappy system costs £14.95 (one nappy, two hush cloths, four paper liners and one microfibre overpant).

Twinkle, Twinkle* Carries a wide selection of nappies including three organic types: Disana organic tie-up for £1.50, Under-the-Nile shaped one size for £9.50 and Ultra-fit (AKA Snug-to-Fit) for £8.75. Twinkle, Twinkle also sell a full range of covers and wraps including a coloured merino wool wrap with poppers or Velcro starting at £14.90. Nappy starter packs cost from £29.50 (for a newborn kit).

Yummies This nappy company carries non-organic prefold nappies as well as the Disana organic nappy set: tie-up nappy (3 for £4.75), brushed cotton liners (3 for £4.25), muslin booster (3 for £5.25) and woollen cover (from £8.99). Yummies also sells nappy accessories like a nappy washbag for £3.45 and washable wipes. They also stock a few organic clothes and a homeopathic birthing kit for £25.99. Order online or by phone. P&p is included in prices. Delivery can take up to 14 days. Phone 01273 672632 www.yummiesnappies.co.uk info@yummies.biz.

• If you want to use a nappy washing service, contact the National Association of Nappy Services to find contact details of one serving your area. Phone 0121 693 4949 www.changeanappy.co.uk info@changeanappy.co.uk.

• Avoid using commercial baby wipes which contain alcohol, strong fragrance, preservative and other chemicals. The Tushies brand offers alcohol-free hypoallergenic wipes with aloe vera (£3.45 for a tub of 80 wipes, £3.19 for a refill), which are useful when travelling, or use washable wipes, reusable cotton muslin, untreated kitchen roll paper, or soak cotton pads in pure water with a couple of drops of lavender essential oil.

Disposable alternatives

If you really need to use disposables at any time—for instance when travelling long distance, during hospital stays or an emergency—there are now seven companies offering 'halfway house' alternatives to the polluting disposables available off the shelf. Some of these use unbleached cores, others are plastic-free. However, bear in mind that many of the plantation forests used for producing these will still be monocultures such as pine which, while renewable, are mostly devoid of native plant and animal life, and replace native forests and land—many international forestry companies are even experimenting with genetically engineered trees which grow faster and bigger.

Moltex Oko Available from Spirit of Nature*, Eco-babes* and Natural Collection*, these are TBT-free unbleached cellulose disposables from Germany in four sizes: standard or big value packs from £12.99 for a pack of 58 mini nappies, and £35.99 for a box of 174. The synthetic gel found in regular disposables is still used. Eco-babes has smaller packs of twelve disposables and six biodegradable nappy bags from £3.25 per pack.

Tushies Imported from the US, these dye-free, gel-free nappies are made from dioxin-free cotton and woodpulp padding. Price start from £8.30 for a bag of 20–40 nappies depending on size, a box of 4 bags starts at £30. Tushies also make alcohol-free wipes with aloe vera. One tub of 80 costs £3.45 and refill of 80 costs £3.19. Widely available from health food stores and mail order companies including Green Baby*, Born* and Natural Child*. www.tushies.co.uk.

Weenees This nappy system from Australia, sold through independent pharmacies and by Lollipop, has reusable waterproof shaped velcro soft pants which can be used with Weenees shaped disposable pads which are flushable (when torn in half and soaked, as they contain no plastic) and compostable (134 days in soil, up to 80 days in earthworm compost and 11 days in septic tank). The nappies still contain polyacrylate absorbent material (super absorber gel) as do regular disposables, but in much lower quantities, and have been shown in tests by an Australian university not to kill earthworms and soil organisms. Weenees are available from Lollipop (01736 799512) and some independent pharmacies and health food stores.

Babywear

At least twelve companies now offer pure organic cotton and wool clothing for babies—ideal if your baby suffers from eczema or sensitive skin. Some include nickel-free snap closures in clothing, and natural buttons made out of wood or ceramics. If you can only afford a few items, choose ones that will be worn closest to your child's skin (i.e. short body suits and vests). Organic babywear is extremely soft, and cotton fleece surprisingly warm.

Beaming Baby* Carries a small range of unbleached organic basics: long and short-sleeved body suits are £4.49, baby t-shirt costs £10.99. The company also has some colourful organic clothes including a turquoise baby dress for £12.99, patchwork dungarees starting at £12.99 and patchwork trousers from £14.99. An organic Gro-bag costs from £22.50.

Bishopston Trading Company Offers fair trade organic clothing for babies, children and adults made in India. Colourful organic baby clothes include appliqué trousers for £13.50, matching appliqué jacket for £15.90. They also have soft baby block toys for £2.95. The company does mail order and has five shops in the UK: Totnes, Bristol, Glastonbury, Stroud, Bradford-on-Avon. Phone 01453 766355 www.bishopstontrading.co.uk info@bishopstontrading.co.uk.

Born* Offers a small range of organic baby basics including a sleep set in natural or blue stripe for £24 (includes sleepsuit, vest, hat and mittens) and a baby bonnet for £5. They are planning to expand their babies and children's range.

Clothworks* This Bath company has a small range of organic cotton clothing in a range of colours: baby hooded top in fleece-backed jersey (£36), stretch waffle top (£12).

Cotton Comfort* Organic clothing and other garments which meet the Oeko-Tex 100 Standard are offered by mail order by Cotton Comfort for those with sensitive skin or suffering from eczema, psoriasis, allergies or burns. Organic mitten pyjamas with feet and mittens attached cost £32 (three sizes). Pyjama leggings with adjustable buttoned shoulder fastenings come in three sizes for £18 (with feet attached); go-over mitten T-shirts (with mittens attached) come in three sizes for £18. Organic baby bodies cost from £8 and a sleepsuit costs £32. Cotton Comfort also sell an organic cot blanket for £22.

Cuddlebabes* This company has a limited range of organic baby clothing including a Green Baby bodysuit for £4.49 and sleepsuits for £10.99. They also sell non-organic products like blankets, slings and shoes.

*Eco-babes** This Essex company has organic cotton garments in sizes for newborns to one year olds. There's a baby body for £4.99, a baby playsuit for £19.99, a sleep suit for £11.99 and a bib for £3.50. Pyjamas come in sizes up to five years old and cost £15. An organic hooded towel costs £9.99.

*Garthenor** A good range of hand-knitted organic pure wool clothing. The baby range includes bootees for £9.50/£10.50, mitts for £7.50/£8.50, jumpers from £28 to £56, leggings from £25 to £43, and hats £9.50/£10.50. The jumpers are available in various styles, including Guernsey and Fair Isle.

Golden Lamb Rugs Fleeces in natural colours from James & Mary Yule's organic farm on the Hereford/Wales border, tanned locally and biodynamically. Ideal for cots, seats and nursery floors, and adaptable for other uses such as car seats. Machine washable with care. Prices from £50. Visitors welcome. Phone 01600 740252 jyulefarm@btopenworld.com.

*Gossypium** Good selection of basics for baby in plain unbleached or printed cotton including newborn babygrow for £14 and sleeping gown starting at £16. Lilac, indigo, unbleached or floral print beanie hat for £6 and jersey knit blanket (75cm square) for £18, Playmat filled with recyled wool for £48 and hooded baby towel for £24.

*Green Baby** Offers a small range of organic baby wear in unbleached natural and colours from low-impact dyes including a grobag for £20.50, babygrow for £9.99, bib and hat for £15.97 and a receiving blanket. Body suits cost from £4.49, booties are £4.99 and scratch mittens are £4.99. The organic cotton fleece lined hemp clothing range comes in navy, red or natural and includes a jacket (£49.99), cargo trousers (£19.99), mittens (£7.99) and booties (£7.99). Also a knitted dress in blue or pink organic Welsh wool (£26.99) and organic corduroy trousers (£12.99).

*Greenfibres** This company offers a large range of basic baby and children's clothing in organic and biodynamic cotton including a long sleeve one-piece in five sizes for £6.70, a woollen baby hat for £5.80, fine socks for £2.90, kimono top for £5.90, pyjamas in four sizes for £17.20, and a short-sleeve one-piece in five sizes for £6.45. A jersey knot baby pouch costs £24.10. Greenfibres also has a range of baby vests, tank tops and leggings in unbleached and printed organic cotton.

Little Earthlings Sells a small range of baby basics in organic cotton including a body suit (£4.49 for short-sleeves, £4.99 for long), a babygrow (£9.99) and a sleepgown (£9.99). They also has an organic sleeping bag for £29.95.Order online or by phone. P& p is free on orders over £100. £5 gift vouchers also available. Phone 028 2954 1214 www.littleearthlings.com info@littleearthlings.com.

For further details about companies marked with an asterisk (*) including how to order, postage & packing charges etc, see the Leading Mail Order Suppliers section on pages 164–167.

Little Green Earthlets Carries the O! Angel range of US-made baby and toddler clothes in organic jerset knit cotton with fruit and vegetable patterns. The garden romper costs £17.50, Veggie patch cardigan (from 9 months) costs £13.95, Grape hat (3 to 9 months) costs £8.95 and the Blueberry trousers (3 to 9 months) costs £12.95. Shop online or by phone, fax or email. P&p costs £1.50 on orders under £10, £2.95 on orders £10 to £50. P&p is free on orders over £50. Phone 01825 873301 www.earthlets.co.uk sales@earthlets.co.uk.

Lollipop Offers organic wool sweaters hand-knitted in Cornwall for ages 1 to 4 years (custom sizes also available) for £34.99. Phone 01736 799512 www.teamlollipop.co.uk enquiries@teamlollipop.co.uk.

Natural Child* Offers three pieces of organic fairtrade baby clothes: long-sleeve body (£5), sleep gown (£10) and babygrow (£11).

Natural Collection* This national mail order catalogue offers a set of pastel baby garments including matching long-sleeve shirt, jacket, leggings, hat and blanket for £27.50.

The Nice Nappy Company* Sells a small range of organic baby basics including a body suit (short-sleeved costs £4.50 and long-sleeved costs £5). They also have Gossypium designed sleepbags (£16) and a kimono (£13) in unbleached or patterned organic jersey knit.

Schmidt Natural Clothing* Schmidt has a large selection of organic cotton and wool babywear, along with other untreated fibre clothing, and offers a starter pack for first time parents containing all the basic items, which can be customized. A colour grown sleeping bag/playsuit costs £20.90. Long sleeved T-shirts for babies in three colours cost from £7.70, and striped woollen bootees in three colours cost £9.80. Other products include a cotton sleepsuit with mittens and feet for babies with eczema from £27.50, separate mittens with ties from £2.50, and a baby romper from £12.30. Demeter biodynamic wool garments are also available.

Smilechild Sells a small selection of organic baby goods including pyjamas for £18, a long sleeved vest for £7 and a playmat for £48. This company also sells non-organic and eco-disposable nappies and toys. Order online, by phone, post, email or fax. P&p costs £3.95 on orders up to £75, over £75 is free. Phone 0800 1956 982 www.smilechild.co.uk customerservices@smilechild.co.uk.

Spirit of Nature* Spirit offers a large range of imported organic and Demeter biodynamic cotton clothes (over forty different summer to winter garments); and unbleached, chemical-free untreated cotton and wool clothing for babies under the Oeko-Tex Standard in its Baby & Children Catalogue. Organic garments in up to four sizes include organic colour grown cotton sleepsuits (£7.90), striped cotton sleepsuits (£12.98) bonnets (£3.90), cotton fleece teddy overalls (£24.90), cotton velvet teddy rompers (£11.90), fashionable jacket (£14.90) and sweatpant (£13.95) two-piece, striped cotton hats with cord ties (£6.89), and leggings (from £8.50). There's also an organic hooded bath wrap for £11.95.

Twinkle, Twinkle* This company sells a small selection of organic sleepwear for baby including a bodysuit in patterned or unbleached (short-sleeved costs £4.49 and long-sleeved costs £4.99), long-sleeved tie body (£4.99) and terry towelling sleep suit (£10.99).

Woolly Caterpillar Run by a mother and daughter team, this company offers 15 pieces of organic baby clothing including a sleep suit for £10.99, long-sleeved body plain or with ties for £4.99 and short-sleeved body plain or with ties for £4.49. They also have an organic wool swaddling blanket for £28.99. P&p costs £3.95. Delivery within 28 days. Phone 01483 278065 www.woollycaterpillar.com info@woollycaterpillar.com.

Toys & accessories

There are many hazards in our babies' and children's toys—both to them and to the environment. They include phthalates or plastic softeners in soft PVC toys, electromagnetic fields from electric-powered toys, synthetic furs and fibres, and the production of plastics used in toys. According to Greenpeace, soft plastic toys made with PVC contain toxic chemicals, which can leak out and be ingested by babies when a toy is squeezed, sucked or chewed. These phthalates have apparently been shown to cause liver, kidney and reproductive problems, and Greenpeace is campaigning for all soft PVC toys to be withdrawn (those 'intended' for the mouth have been). They say that Mothercare, Boots, Woolworths, Toys 'R' Us and the Early Learning Centre, along with other high street shops, sell PVC toys which can contain up to 40% phthalates by weight. Many European countries have taken such products off their shelves, including Sweden, Spain and Italy; Austria banned them in June 1998, and some other European governments are urging retailers not to sell soft PVC toys. Dioxins (some of the most toxic chemicals, which are linked to immune system problems and cancer, and known to be a hormone disrupter) are released into the environment when PVC is made, and some phthalates have been found to be hormone disrupters. While toys made with phthalates often carry the label saying non-toxic, Greenpeace claims that the label on a bottle containing these chemicals for use in a laboratory has safety warnings including: 'may cause cancer', 'possible risk of irreversible effects', 'harmful by inhalation, in contact with skin and if swallowed'. For further information, contact Greenpeace (0800 269 065 www.greenpeace.org.uk). If you must buy plastic goods, choose PVC-free plastics, which have less environmental impact and are less harmful to your baby's health.

Ideally, choose toys made from sustainable wood (never from MDF, as it contains formaldehyde resin) and recycled, unbleached, untreated, chemical-free or organic fabrics. Some of those now available are listed below.

For further details about companies marked with an asterisk (*) including how to order, postage & packing charges etc, see the Leading Mail Order Suppliers section on pages 164–167.

Toys

Born* Sells wide range of wooden toys finished with linseed oil or water-based dyes. Tower with colourful rings costs £9.50. Some toys are handmade in UK Camphill communities from English woods: pull-along waddling duck costs £7.

Cuddlebabes* Sells organic Keptin toys including Rattle cozy (12cm) for £4.49 and Rattle Ring Zmooz with wooden teething ring (14cm) for £4.49.

Dawson & Son Wooden Toys & Games The Dawson toy catalogue contains a variety of wooden toys for use from birth onwards. All are crafted using natural materials from replenishable sources, and non-toxic paints are used. They range from jingle bells from £4.95 and rattles from £3.39, to elaborate pram chains from £13.60, jack-in-the-box toys for £8.50 and activity beads from £19.56. There are even wooden toddler swing seats and a swing, slide and rope ladder wooden set. Order by phone, online or or visit the shop at Winstanley House, Market Hill, Saffron Walden, Essex CB10 1HQ. Phone 08700 367 869 www.dawson-and-son.com info@dawson-and-son.com.

Ethos Offers wooden toys including smile rattle for £3.95. Also organic Keptin toys—their best-selling cuddle zmooz costs £8.95. Phone 0845 0900 184 ethosbaby.com enquiries@ethosbaby.com.

Green Baby* This mail order company and shop in Islington stocks stuffed organic toys including a Teddy bear for £7.99 and a musical stuffed moon mobile for £25.99. Also a good range of wooden toys including stacking tower with coloured rings for £13.99 and a handmade rattle with semi-precious crystals for £10.50.

Greenfibres* Organic cotton toys sold by Greenfibres include an untreated merino lambswool hedgehog (25cm) for £17.90 They sell more organic stuffed toys from their shop in Totnes.

Holz toys Carries wooden toys for babies and children, indoors and outdoors. Little town pram chain with bell costs £16 and birthstone rattle costs £12. Organic soft toys include wool stuffed polar bear for £37.50 and lamb hand puppet for £11. Order online, by phone or email. P&p is £3.95. No p&p on orders over £100. Delivery within 10 working days. Phone 0845 1308691 www.holz-toys.co.uk info@holz-toys.co.uk.

Little Earthlings This company has a small selection of wooden and soft toys. The colourful wooden Moby wriggling fish (suitable from 6 months) costs £7.50 and a dummy holder costs £8.75. Order online or by phone. P&p is free on orders over £100. £5 gift vouchers also available. Phone 028 2954 1214 www.littleearthlings.com info@littleearthlings.com.

Little Green Earthlets Small, basic organic cotton doll (Keptins) from Holland are available from Little Green Earthlets Ltd (as well as a range of wooden toys including a birdy for £2.75, complete Noah's Ark for £21.95 and a pull-along snail for £9.95. Shop online or by phone, fax or email. P&p costs £1.50 on orders under £10, £2.95 on orders £10 to £50. P&p is free on orders over £50. Phone 01825 873301 www.earthlets.co.uk sales@earthlets.co.uk.

Lollipop Offers a range of educational wooden toys including a bag of 50 bricks (£9.99), fire engine (£15.99), weather frog (£2.99) and a milk-tooth holder (£1.99). Phone 01736 799512 www.teamlollipop.co.uk enquiries@teamlollipop.co.uk.

Natural Child* Offers colourful wooden toys including rattles made of rubberwood (toadstool, bell cage or cross for £4.40), a 'first shapes' sorter with colourful shapes to put into holes made of waste wood (£8), and a pram rattle (£7.50).

Natural Collection* This wide-ranging catalogue sells organic cotton stuffed cuddly creatures including a mouse, penguin, frog and dolphin, each for £12.50. Organic colour-grown cotton pig in dungarees costs £26. They also carry wooden toys including a pram chain with triangular colourful figures (46cm) for £14.50. Their original rubber duck is made from natural rubber latex and costs £4.95.

People Tree This UK branch of a Japanese fairtrade company sells an organic cuddle bear stuffed toy for £3 and a beautiful mobile (drop of 36 cm) with different coloured palm-leaf origami fish for £4. Order by phone, post or email from their paper catalogue. P& p cost £3.50 up to £75. Phone 020 7808 7060 www.ptree.co.uk sales@ptree.co.uk.

Schmidt Natural Clothing* Offers a selection of handmade fluffy toys, made of thick organic cotton plush stuffed with sheep's wool. Cotton stuffing is also available. The toys include Teddy Bear (30cm, £30); Hoppy Bunny (23cm, £11) and Polar Bear Hand Puppet (17cm, £17).

Spirit of Nature* Spirit sells a small range of toys including a wooden rattle for £2.95, jingle-jangle pram toy with natural wooden balls for £4.20 and handknitted balls with untreated wool filling for £3.90.

Toyworm Sells handmade and hand painted traditional wooden toys. An abacus with colourful beads costs £2.50, xylophone costs £6.75 and African-made pop-up animals are £2.50 each. Order online or by phone. Phone 0113 2448883 www.toyworm.co.uk.

Woolly Caterpillar This company run by a mother and daughter team sells 123 educational wooden toys including a dancing caterpillar for £12.50 and a frog thermometer (their best selling item) for £2.95. They also have organic cuddly animal toys starting at £5.95. Order online or by phone. P&p costs £3.95. Delivery within 28 days. Phone 01483 278065 www.woollycaterpillar.com info@woollycaterpillar.com.

For further details about companies marked with an asterisk (*) including how to order, postage & packing charges etc, see the Leading Mail Order Suppliers section on pages 164–167.

Slings

Using a sling can make moving around with a young baby very easy, as you become hands-free, and it can be comforting and warm for your child to be carried next to your chest and stomach with good back support.

The Carrying Kind This sling company sells twelve different styles of slings, of which three come in organic fabric. The New Native Baby Carrier, a sling suitable for the first few months, costs £28.50 inc p&p in various sizes. It carries the baby in hammock style, and can be used for older babies in a sitting position. The Baby Bundler, suitable for carrying a newborn to a child of two years of age costs £38 and The Huggababy sling costs £39.95. The Carrying Kind also offers a page on their website for customers to buy and sell second-hand slings. Phone 01992 651717 www.thecarryingkind.com Claire@thecarryingkind.com.

Huggababy Sells Huggababy carrier in conventional and organic cotton. Organic cotton Huggababy with polyester padding costs £39.95, with wool padding it costs £42.95. The company also sells wool sleeping bags, natural lambskins and wooden toys handmade in Wales from native woods including waddling drake duck for £16.95 and whole nursery ark for £76.95. Order by phone, fax, post or online. Orders dispatches between 1 and 3 days. Prices include p&p. Phone 01874 711 629 www.huggababy.co.uk.

FAMILY GEAR

Family clothing

There are a growing number of companies now offering pure organic cotton, linen and wool, and chemical-free clothing for older children and for the whole family.

Bishopston Trading Company A large range of organic fairtrade colourful clothing for men, women and children is available from this Fairtrade co-operative, which has five shops in the West of England, and also sells by mail order. The co-op was set up in 1985 to create employment in the south Indian village of KV Kuppam, to where all profits are sent. Phone 01453 766355 www.bishopstontrading.co.uk info@bishopstontrading.co.uk.

*Cotton Comfort** Organic clothing and other garments which meet the Oeko-Tex 100 Standard are offered by mail order by Cotton Comfort for those with sensitive skin or suffering from eczema, psoriasis, allergies or burns.

*Garthenor** Organic, hand-knitted pure wool hats and jumpers for adults and children. You can choose the yarn from which the garment is knitted. All natural colours—the wool is neither dyed nor bleached.

*Gossypium** A trendy selection of clothing for adults, children and babies. Yoga wear and accessories also available. Phone 01273 897509 (info), 0800 085

6549 (mail order) www.gossypium.co.uk customercare@gossypium.co.uk or catalogue@gossypium.co.uk (for free catalogue).

*Greenfibres** Greenfibres offers one of the widest ranges of family clothing in organic and natural fibres from toddlers to adults in its full colour catalogue. Organic cotton vests, briefs, long johns, socks, pyjamas, night shirts and boxer shorts are available, as well as organic cotton lingerie, and chemical-free Harris Tweed with organic linen waistcoats, trousers and skirts and an untreated 100% cashmere scarf. Greenfibres also sells bourette silk garments and organic silk underwear, hemp shirts, trousers and jackets.

*The Healthy House** This company, which specializes in products for people with allergies, chemical sensitivity, asthma and eczema, has a small range of organic clothing for women including underwear and tops.

*Natural Collection** Offers a small range of stylish organic clothing for adults including underwear, pyjamas, bathrobes and exercise and yoga wear.

Natural Instincts Offers a range of Irish organic woollen and cotton clothing for children and adults, as well as household textiles. P&p is dependent on the weight of the garments. Phone 00353 733 8256 or 00353 733 9064 doogan@iol.ie.

Patagonia The first active wear company to use 100% organic cotton, this US company has over 60 stockists in the UK. They also sell fleeces made from recycled soda bottles. There are adults' tank tops, T-shirts, polo shirts and shorts, smart trousers and denim jeans, sports dresses and a small range of children's wear in organic cotton. Patagonia has spring/summer and autumn/winter catalogues, with products available by mail order: for a catalogue (which also lists UK outlets, as does their website) phone 0033 1 41 10 18 18 (European head office in France) www.patagonia.com euro_hotline@patagonia.com.

People Tree Fashionable organic and fair trade clothing for men, women and children. Order by phone, post or email from their paper catalogue. P& p cost £3.50 up to £75. Phone 020 7808 7060 www.ptree.co.uk sales@ptree.co.uk.

*Schmidt Natural Clothing** The range includes organic pyjamas, a polo neck jumper and a chemical-free woollen cardigan for children, women's bras (including nursing bras) and breast pads in organic cotton. Other adult clothes include nightdresses, silk tops, leggings, slips and skirts.

*Spirit of Nature** Offers one of the largest, most fun and most professionally presented range of imported organic and Demeter biodynamic cotton clothes (over forty different summer to winter garments), and unbleached, chemical-free untreated cotton and wool clothing for toddlers in its Baby & Children Catalogue, produced to the Oeko-Tex Standard.

Bedding

Many UK companies now offer organic bedding (mattresses, cots, beds, pillows and pillowcases, sheets, duvets and duvet covers, and sleeping bags) which has not been treated with flame-retardant chemicals (which have been linked to cot death), formaldehyde, strong bleaches, dyes (which can contain heavy metals), or other chemicals. We spend about a third of our lives in bed—so why not sleep in natural, chemical-free fabrics, rather than breathe in chemicals throughout the night? If you're on a tight budget, at least make sure that the bedding closest to your baby is organic or untreated, and gradually replace old bedding with these items as well. If you do buy a new treated mattress or base, allow these to outgas away from adults and children for several days before covering and using.

Beaming Baby* Organic cot sheets start at £4.

Born* Stocks Stokke beech furniture that converts for use as child grows, including cots and chairs.

Clothworks* Sells organic cotton sheets and duvet covers for cots. Fitted and flat cot sheets cost £16 each, cot duvet covers cost £27 each, and blankets £25.

Eco-babes* Eco-babes now offers organic cot bedding including fitted cot sheets from £11.99, cotton fleece blanket for £19.99 and an organic bed protector for £14.99.

Gossypium* Organic unbleached bedding including a flat sheet for £24, a fitted sheet for £25 and duvet cover for £44. Organic towels and table linens are also available.

Green Baby* This mail order company and shop in Islington sells naturally oiled beech cots starting from £129. An organic bed protector costs from £6.99 and their organic cotton bumper costs £34.99. Bedding includes organic cotton blue bunny print duvet and pillow case set for £34.99, blanket with bunny print trim for £20.99 and fitted sheets in woven, jersey and terry starting from £12.99 They also have organic pillows and duvets as well as biodynamic wool baby blankets in blue, red and natural for £24.99.

Greenfibres* Greenfibres has the largest range of organic and chemical-free bedding in the UK, including mattresses and beds. Their enormously comfortable child and adult-sized mattresses are filled with organic latex, organic wool and organic cotton; another one contains organic wool, organic cotton and coir. All mattresses comply with UK fire regulations as wool is naturally fire-retardant and needs no chemical treatment. Mattress prices start at £129 for children's, and £325 for a light density single adult mattress. For £10, Greenfibres can send you a sample of the mattresses for 14 days. They also sell two styles of metal-free beds made from solid beech or cherry. Adult's bed prices start at £680. The Merlin cot/bed is made from European solid beech from sustainably managed forests and treated with natural oils. It converts easily to a child's bed and fits the child's latex mattress. The organic Merlin bedding range includes a crib drape (£49) pillowcase, duvet cover and fitted sheet (£46) and cotton velour blanket (£47) in a colourful pattern. Merlin organic cotton patterned fabric is also available for £19 per metre.

An organic cotton mattress protector costs from £15.90 and a naturally waterproof sheet costs from £18.20. Duvets are filled with organic cotton, organic linen and/or organic wool in adult and children's sizes. Adult-sized duvets can also be filled with raw silk or camel hair. Child's duvets (which fit the Merlin cot/bed) start at £67 and adult sizes start at £97. Organic cotton and lambswool mattress covers are also available. Greenfibres also offers a number of organic wool, organic cotton, cashmere, and alpaca blankets, including an unbelievably soft organic cotton fleece blanket (150x200cm) for £84. Sheets, both fitted and flat, as well as pillowcases and duvet covers are available in classic unbleached, unbleached woven stripes, yellow check and colour grown stripe, sold individually or as sets. Sheet sets start at £55.80 for the single size. Pillows (regular, bolster, Euro square and neck roll sizes) filled with organic wool, spelt, horsehair or millet and covered in organic cotton start at £21.

*The Healthy House** Not organic, but untreated cotton pillowcases, sheets, and duvet covers. Also lightweight organic cotton duvets (£120-£225 plus £7.50 p&p) and organic heavyweight cotton duvets (£150-£295 plus £7.50 p&p) are among the hypoallergenic and natural bedding products sold by this mail order service. Dust mite proofing (wraps or a non-chemical treatment) is available. If you can't afford an organic mattress, consider an untreated cotton one (from £78 for cot size with cotton and polyester filling).

Hippychick Carries cot blanket in 10 colours including Oeko-tex approved cotton fleece blankets for both cot and pram and an organic pram blanket (75x100cm) for £22.50. Also sells AromaKids range and leather baby shoes. Phone 01278 434440 www.hippychickltd.co.uk.

Lollipop Sells an organic wool cot blanket for £44.95 and an organic wool pram blanket for £24.95. Also an organic lambswool for £59.99. Phone 01736 799512 www.teamlollipop.co.uk enquiries@teamlollipop.co.uk.

*Natural Collection** This nationwide mail order catalogue has organic cotton bedding in, natural (plain or brushed), white (non-chlorine bleached) with blue stripe or woven stripe with waffle edging Pillowcases are available from £13.95 a pair; single, double and king size flat and fitted sheets range from £34.95 to £49.95; and duvet covers in three sizes range from £39.95 to £59.95. Organic cotton and wool filled duvets start at £120. Untreated wool pillow, and neck pillow, an organic millet pillow for £49 and cotton and wool mattress pads are also available. Zip up, fleece lined organic cotton baby pouches with cute bunny print cost £28.50. A checked blanket (2x1.5m) of UK produced organic wool and dyed with pink and green plant dyes is £99.50.

Natural Mat Sells chemical-free handmade mattress for cots and cribs. Cot mattress contain natural latex, wool, horsehair and coir wrapped in an unbleached cotton herringbone ticking. The mattresses are naturally fire-retardant (complying fully with UK regulations) and are dust-mite resistant. They have three standard sizes: 56x118cm, 60x120cm and 70x140cm as well as oval. Also they can custom make any size. Price start at £90. Natural Mat also sells a crib/Moses basket woven from Somerset willow starting at £115 and a convertible wooden cot made from pine or maple and walnut for £1050. Order direct from the company online, by phone or email. P&p costs from £4.50. Order processed within 3–5 working days. Phone 020 7689 www.naturalmat.com info@naturalmat.com.

Naturally Nappies This company offers an organic woollen sleeping bag in red or blue for £11.50. They are planning to stock natural wooden cots and baby hammocks. Order by phone or email. Phone 0870 745 6641 www.naturallynappiesuk.co.uk orders at naturallynappies.com.

Organic Sheepskins Based in Hereford this company produces biodynamically tanned sheepskins (and goat skins) from organically reared animals. They come in a range of natural colours and naps owing to the different breeds. Prices start at £45. Delivery costs £5 and is within 14 days. Phone 01989 730615 www.organicsheepskins.co.uk enquiries@organicsheepskins.co.uk.

The Organic Wool Company This UK company sells a natural coloured baby blanket in two sizes: small (120x90cm) costs £40 and large (180x120cm) costs £65. Also sells naturally plant-dyed and rare breed woollen blankets, scarves, rugs and throws. P&p cost £3.50 on orders up to £150, p&p free over £150. Phone 01239 821171 www.organicwool.co.uk douglaswhitelaw@organicwool.freeserve.co.uk.

People Tree Organic Jersey bed linens: single fitted sheet costs £32, single duvet cover costs £25 and pillow case costs £10. Order by phone, post or email from their paper catalogue. P& p cost £3.50 up to £75. Phone 020 7808 7060 www.ptree.co.uk sales@ptree.co.uk.

Schmidt Natural Clothing* Organic bedding on offer from Schmidt includes handy swaddling cloths for £4.90 (small blankets), organic cotton flat or fitted sheets and duvet covers (£19 and £29.90 for baby, or (from £30 to £70) for adults, pillowcases (£14 a pair), an organic wool sleeping bag (from £21), an organic wool blanket (£19.50) for babies and children, organic wool mattresses (price on request), and an organic sheepskin (biodynamically tanned) made in Somerset for £69.50. There are also organic and wild herb pillows for £14.70.

Spirit of Nature* Spirit offers a ladybird organic cotton fleece blanket for £39.90 and coir and anti-dustmite treated lambswool cot mattresses (naturally flame retardant with no chemicals). They also sell a protective brushed cotton and natural rubber mat for mattresses.

Willey Winkle Willey Winkle of Hereford offers traditional craftsman-made organic cotton and wool bedding for babies and children, including organic mattresses for cots, cribs, Moses baskets and children's beds. The wool used, which is hand-washed and scoured prior to stuffing, is from sheep raised on the organic Highgrove Estate at Gloucestershire. Cotton used is organic, unbleached herringbone ticking. Prices range from £95 for an organic Moses mattress to £269 for an organic cot mattress and £337 for a junior bed mattress. Adult size mattresses are available from £818 up to £1630 for a king size mattress. Organic outer covers are available to fit cot, Moses and crib mattresses from £48. Organic pillows are £68. P&p costs from £13.99. Phone 01432 268 018. www.willeywinkle.co.uk.

Bathwear

Several companies now offer ranges of soft, luxurious organic bathwear such as flannels, towels and bathrobes.

Greenfibres* Greenfibres sells three types of unisex robes: natural organic cotton terry bathrobe (£49.50) and organic cotton jersey house coat (£42). Other organic cotton products for the bathroom include bath, shower and hand towels and wash mitts in unbleached organic cotton and organic linen towelling. A linen towel set (1 bath, 1 shower, 2 hand and 2 mitt) costs £129, while the organic cotton towel set (same components) costs £59. They also have an organic cotton hair wrap towel for £15.80. Wash cloths are also available in a pack of four in natural colour for £10.

Natural Collection* Organic cotton waffle weave bathrobes (£55), towels (£12.95 for hand, £22.50 for bath) and slippers (£7.95) are available in organic cotton (blue or ecru) from this nationwide mail order catalogue. A terry towelling unisex bathrobe comes in ecru, red or navy for £55. Their bath (£23.50), shower (£13.50), hand towels (£9.50) and face cloths (£6 for 3) come in seven colours.

Natural Instincts Offers adults organic cotton bathrobes for £59.95 and a small range of organic towels. P&p is dependent on the weight of the garments. Phone 003 53 733 8256 or 003 53 733 9064 doogan@iol.ie.

Spirit of Nature* Has a hooded organic cotton bathwrap for baby for £9.95 and an organic cotton bathrobe for mum for £49.56.

Fabrics

Consider making your own bedding and/or clothes from organic fabrics, if you have the time and expertise (or go on a course to gain it).

Greenfibres* Greenfibres has organic cotton in flannel, cretonne and denim available from £4.80 a metre. Also available is undyed organic linen from £19.90 a metre. Hemp, silk and organic wool are also available by the metre. For swatch cards and information on fabrics send £3. Organic knitting wool is also available in 2, 3 and 5 ply in natural and 16 hand-dyed plant colours. Natural is £6.50 per 100 grams or £55 per kilo, and dyed is £8.00 per 100 grams or £70 a kilo. For samples send £2. Organic cotton knitting yarn also available.

The Organic Wool Company Sells organic knitting wool in the natural greys and browns of Jacob and Castlemilk Morri sheep. One pack of ten 50g balls costs £29.95. P&p cost £3.50 on orders up to £150, p&p free over £150. Phone 01239 821171 www.organicwool.co.uk douglaswhitelaw@organicwool.freeserve.co.uk.

Texture This small shop in Stoke Newington sell made-to-measure curtains and roman blinds out of organic fabric. Texture also sells organic fabric, organic bedding and Auro paints. Phone 020 7241 0990 www.textilesfromnature.com jag@textilesfromnature.com.

Other handy goods

Apollo A convertible push-chair has been designed with air filtration to protect babies from pollutants caused by traffic on busy streets. Contact Babysphere for information about the Apollo systems. Phone 0870 606 0800.

Green Baby* This one-stop baby shop and mail order company sells glass bottles with silicon or latex teats (certified free of plasticizers and other chemicals that can leach from plastic bottles) starting at £2.35. They also sell PVC and nickel-free baby cutlery and a PVC-free no-spill cup.

Greenfibres* Greenfibres sells an organic cotton string shopping bag for £3.90 and an organic logo-embroidered canvas bag for £6.50. It also has breastfeeding pads in untreated wool or raw silk costing £3.90.

Schmidt Natural Clothing* Organic cotton nursing pads for breastfeeding mothers for £2.75 a pair, or a pack of three for £8.50, hairbrushes, toothbrushes, and more.

Organic pet food & alternative treatments

If you have pets, look out for organic pet food brands in healthfood stores, supermarkets, or by mail order; and consider natural alternatives to chemical pet treatments in the home, which can be toxic to both pets and humans (especially babies). There are also now some vets using homeopathic and other natural remedies—shop around.

Denes (natural pet food and herbal remedies) Phone 012273 325364 www.denes.com info@denes.com.

Pascoes Organic Pet Food Available from Graig Farm mail order and outlets. Cat food costs £1.28 for 375g. Dog food costs £4.02 for 1.5kg. Contact Pascoes customer services for more information. Phone 01278 444555 www.pascoes.co.uk info@pascoes.co.uk.

Think Natural Sells natural pet remedies including supplements and homeopathy. Phone 0845 601 1948 www.thinknatural.com.

Yarrah Organic dog and cat food available from some health food stores and Simply Organic*. Phone 0800 34443888 www.yarrah.com info@yarrah.com.

For further details about companies marked with an asterisk (*) including how to order, postage & packing charges etc, see the Leading Mail Order Suppliers section on pages 164–167.

Bodycare & toiletries

You can refuse the Bounty Bag, which is given to all new mothers by salespeople just hours after the birth of their babies. The bag includes conventional products containing many unnecessary chemicals and additives (mother and baby bodycare, disposables, perfumed nappy bags and baby wipes), together with additional sales literature and offers to encourage you to buy more of these products. Bounty pays hospitals to be allowed to do this, the price per mother varying between 70p and £1 depending on the nature of the contract signed. Several hospitals have refused to use Bounty.

Bounty packs and guides distributed in hospitals no longer carry direct advertising for artificial formula, bottle feeding or breastmilk substitutes. However, Bounty is still involved in the promotion of artificial feeding: it sells part of its antenatal mailing list to formula companies so they can market directly to mothers, while the 'Progress' pack, which mothers can pick up once they return home from hospital, has large amounts of promotion for follow-up formula and other breastmilk substitutes.

There are now many safe organic, environmentally friendly and chemical-free alternatives for mother and baby and the rest of the family, such as deodorants, tampons, toothpastes, soaps, shampoos, moisturisers, body lotions and essential oils. Read the labels of the ingredients in those toiletries and you may be in for a shock—especially ones aimed at your precious newborn, which can contain many chemicals, detergents and synthetic fragrances. Why eat organic food to avoid all those artificial additives and chemicals going into our bodies, and then smother or spray ourselves all over with substances which can include highly flammable petrochemical solvents such as butane and propane, aluminium chlorohydrate (an aluminium salt to avoid), titanium dioxide (a white pigment used in paint and ordinary toothpaste, which is a suspected animal carcinogen and environmental pollutant), the laboratory preservative formaldehyde, pesticide residues from treated plant extracts, artificial sweeteners (in children's toothpastes), sodium lauryl sulphate or SLS (a skin-irritating foaming agent), and many other harmful ingredients? Why on earth use a deodorant such as an anti-perspirant on sale in Boots, whose label carries numerous warnings on it, including 'extremely flammable' and 'do not spray near eyes or face'? Not the sort of product you want to have around small children.

We absorb the chemicals in conventional products through our skin, which is considered to be the largest organ in the body, absorbing, filtering and eliminating waste products. So perhaps we shouldn't smother it in anything that we wouldn't be prepared to eat! Some experts believe that up to 60% of what goes on to our skin is absorbed—up to two kilograms of 'muck' from lotions, make-up, sunscreens and other toiletries a year. And a baby's skin is up to five times thinner than adult's. Then there are toothpastes and oral hygiene products we actually put in our mouths, hair care, and the sanitary products we use on intimate parts of our body (which can contain dioxins and other undesirables).

Pregnant women, babies and small children are especially sensitive to artificial perfumes, additives and chemicals used in these goods, and these can lead to bouts of eczema, headaches and allergies. Many pregnant women actually find they are repulsed by the smell of many ordinary toiletries—your baby's way of ensuring you steer clear of them! And as with residues in food, the combined and cumulative effect of these chemicals is worrying. There's also the production methods to take into account, and their effect on the environment. The products listed use organic ingredients or are certified organic, contain little or no animal ingredients, and are not tested on animals. Many of the companies making these goods recycle and use green technology, and one uses only green energy sources such as wind.

Organic standards and regulations for 'health and beauty' products were launched in April 2002 by the Soil Association, and UK organic companies such as Neal's Yard Remedies, Hambleden Herbs, Weleda and The Green People Company are already leading the way with certified ingredients and products. They will be able to offer professional guidance, put you in touch with organisations or individuals who run aromatherapy and other relevant classes, and give you details of interesting relevant publications. If you're really keen, you can also make your own cosmetics and bodycare products: Neal's Yard Remedies can provide you with a book on that very subject.

Using skin and bodycare products made with organic ingredients can give you a wonderful lift as your skin 'drinks in' the pure oils and herbs, and the waft of aromatic oils heightens your senses and lightens your mood—many of the products are truly luxurious.

Of course, we don't really need many bodycare products if we're putting good foods and oils into our bodies in the first place (for example, dry skin can result from a lack of essential fatty acids and vitamin E in our diet). Nutritionists and naturopaths will tell you that the best way to improve skin condition is to work from the inside out rather than the other way around. And babies especially don't need a lot of toiletries—or any, if you're a purist. Your baby can be bathed once every few days instead of every day, as young babies don't tend to get too dirty, and can be sponged if necessary; and you can either not wash their hair with shampoo at all (perhaps just rinse it), or do so seldomly. You can add one or two drops of appropriate essential oil such as lavender to their bathwater, especially after taking your baby to a chlorinated swimming pool. They require very little soap, as it can dry out their skin, and they can certainly live without bubble baths, powders, lotions and moisturisers; even nappy rash can be treated using a silk liner, with raw egg white, or with airing. A little massaging all over with pure organic almond oil or similar every now and then, however, can be beneficial and bonding for both parent and baby. And if you have dry skin and you don't have any moisturisers to hand, you could reach into your fridge and use some of your essential fatty acid oil blend!

Over 40 brands or companies offering certified organic bodycare toiletries, or those made with organic ingredients, are listed, along with hair care treatments which don't contain strong artificial chemicals. Also consider using recycled toilet paper, which is either unbleached or treated with oxygen bleach rather than dioxin-producing chlorine bleach.

Absolute Aromas Over 100 essential oils and 22 carrier oils are available from Absolute Aromas. They have 21 certified organic essential oils ranging in price from £3.99 for Eucalyptus to £9.99 for Rose Otto. They do five organic carrier oils ranging from £3.99 for Sunflower oil to £8.99 for Jojoba. The range is available from shops and by mail order from Revital*. Contact company directly for stockists. Phone 01420 540 400 www.absolute-aromas.co.uk relax@absolute-aromas.com.

Active Birth Organic personal care products from this educational organisation include Pregnancy Massage oil with neroli, petitgrain and sweet orange (100ml for £14.95), Organic Unscented Massage oil (30ml for £9.95), Pregnancy Bath oil (100ml for £14.95), Labour Massage oil with lavender, sweet marjoram, clary sage and geranium (30ml for £12.95). They also have an organic Postnatal Bath oil (100ml for £14.95) and a Postnatal Herb Bath with loose herbs including marigold, comfrey and St. John's wort leaf (100g packet for £9.95). Their non-organic natural range includes ultra-rich anti-stretch mark cream, nipple care cream, nappy gel, baby lotion and baby cleanse. They also carry Bambino Mio nappy components. Many products include 'buy one get one free' offers. Orders over £50 receive a 15% discount. Order by telephone, fax or online. P&p is £3.95. Phone 020 7281 6760 www.activebirthcentre.com.

Aroma Kids This company began with 4 pre-diluted massage oils specially formulated for children, and has since expanded the range to include toiletries suitable for babies and children. It is NOT organic, but based on natural ingredients. Ellie Smellie barrier cream costs £3.89 for 100ml, Ellie Smellie baby wipes, £3.90 for a tub of 50, Joob-Joobs body wash, £3.99 for 100ml, Crabby Chris massage oil with mandarin, camomile and lavender, £3.99 for 30ml and Noo-Noos newborn massage oil, £3.99 for 30ml. Toby Tumble balm with lemon, lavender and tea tree for scrapes costs £3.29 for 25ml. Aroma Kids products can be found in shops and mail order companies or order online direct. Phone 01278 671461 www.aromakids.com info@hippychick.co.uk.

Avalon Organic Botanicals 50% organic ingredients are used in this imported range of lotions, soaps, deodorants, washes, shampoos and conditioners (£4.99), found in many leading independent organic supermarkets and from Revital*. Phone 01932 334501 www.avalonnaturalproducts.com (US site).

Aveda Hundreds of up-market bodycare products using organic ingredients are available from this US brand of toiletries. Found in many department stores and special Aveda Lifestyle stores, products include the 100% organic All Sensitive Body Formula for £16, and Rosemary and Mint Shampoo for £7.50. The Aveda Institute with the Environmental Lifestyle Store which showcases all products is at 174 High Holborn, Covent Garden, or find Aveda products at Selfridges, Harrods, Harvey Nichols, etc. Phone 01730 232 380 (customer services).

Beaming Baby Offers an organic lavender baby bath range including Bubble bath (300ml for £3.99), Shampoo and Bodywash (300ml for £3.99), Lotion (200ml for £3.49), Baby Oil (200ml for £4.99) and Nappy Cream (100ml for £3.99). Beaming Baby also sell baby body care products from Green People*, Natalia, Weleda and Natracare.

BiOrganic Hair Therapy A mail order catalogue of hair treatments without harsh chemicals—instead they contain herbs, essential oils, vitamins and minerals. The company, which has been going for fifteen years, is in the process of tracing suppliers throughout the world to get ingredients organically grown. P&p is 60p per item. Phone 01384 877951 www.biorganic.co.uk.

Bo Weevil ECOTTON brand non-chlorine bleached organic cotton pad rounds from the Netherlands are available through independent healthfood stores (£1.29 for 50). Also available as pleated organic cotton wool (50g for £1.19).

Circaroma This new company has a large range of sumptuous aromatherapy organic creams, massage oils, floral waters, balms and essential oils. The Absolutely Organic range uses 100% organic ingredients and includes a calming body oil, 100ml for £12.50 and facial oil, 25ml for £10. The hand & body cream with organic cocoa butter, shea butter and lavender is especially luxurious (120g for £9.50). The children's cream contains calendula tincture, carrot oil and cocoa butter and no essential oils, so is suitable for delicate skin, 50g for £5.50. Circaroma also sell ten certified organic essential oils ranging in price from £3.75 to £5.00 for 5ml. Six and twelve bottle boxes available. Candles are made from beeswax and vegetable wax and come in jars in 3 sizes (£4.50 for small). Gift wrapping is available. Order direct from Circaroma. P&p is £3.00. Orders dispatched within ten days. Phone 020 7249 9392 www.circaroma.com sales@circaroma.com.

Deodorant Stone/Deo-Crystal Non-toxic and hypoallergenic, the crystal deodorant stones are made from natural mineral salts without harmful aluminium chlorohydrate (found in anti-perspirant roll-ons and sprays), perfumes, emulsifiers or propellants. Breastfeeding or nuzzling babies can consume your deodorant residues. Simply wet the stone and rub on generously under arms. Each 60g stone lasts about six months, and a 120g stone and liquid roll-on are available. Stones are available from The Green Shop*, Xynergy Health Products*, from Neal's Yard Remedies*, Cuddlebabes*, Natural Collection* and Cotton Comfort.

Desert Essence Organic Australian tea tree oil and some other organic ingredients are used in this extensive range of toiletries from the US, which includes a babycare range (baby lotion, 240ml for £4.95, baby powder,125g for £4.95 and bubble bath, 240ml for £4.95) liquid soaps, toothpastes, deodorants, shampoos, conditioners, lip balm and organic essential oils (15ml for £4.45). They are now widely available from healthfood stores and by mail order through Revital*.

Earth Mother Two Muswell Hill aromatherapists have created aromatherapy massage and relaxation oils and creams using certified organic vegan ingredients specifically for pregnancy, labour, post-natal recovery, and for babies and children.

For further details about companies marked with an asterisk (*) including how to order, postage & packing charges etc, see the Leading Mail Order Suppliers section on pages 164–167.

There's a Tummy Stretchmark Cream for £8.50, Labour Massage Oil with lavender and neroli for £7.50, a Labour Kit with two massage oils, Lavenderflower water, and Five Flower Remedy for £17, a Baby Massage Oil for £6.50 and a Soothing Nappy Gel for £6.50. Postnatal and children's products are available too, as are Little Miracles flower remedies. Order by phone or post; only cheques accepted; p&p is £1.50 on one item, £3.50 up to £50 and 10% over £50. Phone 020 8442 1704 www.houseofshen.co.uk info@houseofshen.co.uk.

Earthbound A UK company set up by young mother Jo Ordonez-Sampson using organic ingredients such as carrot and calendula oils, herbs such as arnica, rosemary, oatmeal, lavender, and aduki beans and sunflower oil, to make a range of toiletries. Beeswax is used as a preservative. Products include 100% organic Baby's Cream (£6.05), Baby Oil (£6.95) and Baby Soap (£3.85). For adults they have Facial Scrub, Relaxing Oil, Anti-cellulite Massage Oil, Shea and Sandalwood Body Butter and Carrot Face Cream. Creams cost from £7.25for 60ml pots. Also make custom tummy oils. Order online or by phone or email. P&p £1.80 for up to five products, £2.60 for over five. Phone 01597 851157 www.earthbound.co.uk sales@earthbound.co.uk.

Earth Friendly Baby A popular baby toiletry range made in the US. Although not organic, it is claimed to be all natural. The range includes Lavender Cleansing Bar for £2.49, Calendula Day Cream (43g) for £2.99, Red Clover Nappy Cream (43g) for £2.99, Camomile Shampoo and Bodywash (251ml) for £3.99—very popular with parents. A starter kit with testers all four products costs £3.99. Available from health food stores and many mail order companies. Phone 020 8206 2066 (their UK distributor, Health Quest) www.earthfriendlybaby.com (US site) info@healthquest.co.uk.

*The Green People Company** This range of cosmetics and toiletries—including new organic baby and child ranges—uses organic and natural ingredients including herb and plant extracts, vegetable oils and essential oils. Their 100% organic products are certified by the Soil Association. Products include organic toothpastes which are free from fluoride, sodium lauryl sulphate and artificial additives and sweeteners (£2.99 for 50ml or a two pack for £4.99) such as Spearmint & Aloe Vera, Fennel, Citrus, Mint and Happy Kids (mandarin flavour). Ingredients include calcium carbonate, carageenan (seaweed), grapefruit seed extract, and herbs and fruits. There are also aloe vera and rosemary shampoos, and conditioners (from £5.49 for 200ml), sunscreens for children and adults (from £11.99 for 150ml) which do contain titanium dioxide, and many other beauty products. The Organic Baby range includes 100% organic Baby Salve (75ml for £6.45), Baby Wash (100ml for £4.25) and Baby Lotion (100ml for £6.45). The Organic Children's range includes Shower Gel (150ml for £4.49) and Shampoo and Conditioner (150ml for £4.49) Full ranges can be seen at leading healthfood stores, and is available by mail order from Green People*.

Hambleden Herbs The Organic Herb Trading Company's Hambleden Herbs brand has one of the UK's largest ranges of organic herbs and herbal products, using many herbs from its own gardens. Organic herbal tinctures from Hambleden (a 50ml tincture costs from £5.90) can be added to baths. There's also organic Red Henna for colouring and conditioning hair (£2.90 for 200g), Organic Almond Oil for massaging (£6 for 100ml or £15 for 500ml), Organic Lavender Water and Roman Chamomile Water (£4.90 for 100ml). Available through leading health stores.

Dr Hauschka Biodynamic herbs and ingredients are used in this full range of imported cosmetics and bodycare which has been given the thumbs up by national beauty writers. Cleansing Cream costs from £10 for 50ml; Moisturising Day Cream from £12 for 30ml; Cleansing Clay Mask from £15 for a refill for a 90g jar; Neem Nail Oil £18 for 30ml; and Rosemary Food Balm £10 for 30ml. For an ideal introduction to the range try sample tin kits, which are also useful when travelling: the Daily Bodycare Kit costs £10 for eight items, and a Daily Facecare Kit with six items costs £10. An extensive mail order catalogue is available from UK distributor Elysia Natural Skin Care, or see the full range at many health food stores including Fresh & Wild, Planet Organic and As Nature Intended. Phone 01386 792 622 (Elysia mail order) www.drhauschka.co.uk enquiries@drhauschka.co.uk

Herbs Hands Healing A large range of custom made bodycare and home care products, tinctures, capsules and ointments using organic and wild ingredients are available by mail order from Herbs Hands Healing. When you call, let them know if you are pregnant, breastfeeding, or planning to conceive. Phone 01379 608 007 or 0845 345 3727 www.herbshandshealing.co.uk.

House of Shen A range of products for use from pregnancy onwards. Their Babycare Kit (£27.00) contains Massage Oil, Soothing Nappy Gel, Baby Lotion, Baby Bath Hydrolat and Chamomilla. Phone 020 8442 1702. www.houseofshen.co.uk.

In Harmony Aloe & Cucumber roll-on deodorant uses the crystal deodorant stone formula. It costs £4.95 from Xynergy Health Products*. P&p on total orders is £2.95. Body spray (purely pump action) in Unscented, Herbal or Aloe costs £4.

Jurlique Organic and biodynamic ingredients are used in this Australian bodycare range for adults and babies. The large range includes massage oils, hair care products, moisturisers, lipsticks and cream masks. The new baby range includes Baby's Barrier Cream (125ml for £20.95), Baby's Calming Bath Oil (100ml for £10.45), Baby's Massage Oil—Calming and Colic Relief (100ml for £16.55 and 50ml for £13), Baby's Gentle Cleansing Balm (100ml for £13.60), Baby's Shampoo and Bodywash (125ml for £8.95), Baby's Soothing Bubble Bath (125ml for £10.45) and Baby's Soothing Cream (125ml for £20.95). For adults. Rosewater Freshener is £18.95 for 100ml, Wrinkle Softener Cream is £17.35 for 8ml and Body Care Lotion is £22.50 for 100ml. Jurlique products are available by mail order from The Naturopathic Health & Beauty Company with an added £2.50 p&p; and from some health food stores including Fresh & Wild and Olivers Wholefoods. Phone 0208 841 6644 www.jurlique.com (US site) sales@nhb-company.co.uk.

Kinetic Organic Body Polish Slightly exfoliating body scrub with organic ingredients. One 740ml tub costs £4.50 and comes in calming, clearing and revitalising varieties. Available from Revital* and health food shops.

Lavera German-made range of products using organic and wildcrafted plant ingredients, recommended by Oko-test magazine over 80 times. The Baby range includes Baby & Children Almond Cream for £5.95, Baby Bath Oil Almond for £6.80, Baby & Children Shampoo Almond for £5.50, Baby Oil Almond for £6.80. The many products for adults include Jojoba & Aloe Vera Body Lotion for £5.55, suncreams and chemical-free cosmetics. Products available at some independent health food stores. Lavera UK is based in Scotland. Order online or by phone. P&p costs £2 minimum. Phone 01557 814941 www.lavera.co.uk.

Lindos This imported range of baby bodycare products made by Walter Rau (established forty years ago) contains biodynamically grown plant extracts 'as far as is practically possible'. Some products do contain animal ingredients such as tallowate (animal fat) and lanolin (grease from wool). The range includes camomile Baby Soap (£1.50-£2.20 for 100g), camomile Protective Nappy Cream (£3-£4.30 for 50ml), Baby Moisturising Cream (from £3 for 50ml), Camomile Baby Oil (from £5 for 200ml), Baby Bath Milk (from £5 for 200ml) and Baby Shampoo (from £5 for 200ml). There's also baby powder, using maize flour. A mail order catalogue of the full range is available from Elysia Skin Care (phone 01527 832 863; sample sizes are free), or by mail order from Elysia*. Some Lindos products can be found in leading healthfood stores.

Logona Logona (from Germany) has been producing natural cosmetics for more than 18 years, and now has a large range of adult, baby and children's bodycare products using organic ingredients such as vegetable oils and organic herbs 'whenever available'; otherwise wildcrafted herbs (grown in the wild, but not registered as organic) are used. Products include surfactant-free shampoo, Lip Balm, Scrub, Baby Bodycare Oil (£5.04), Baby Cream Bath (£4.79), Baby Protective Cream (£3.96), and an Aloe & Green Tea range. There's also Baby Herbal Tea, Kids Tea (with soothing herbs), Kids Face & Body Lotion (with natural UV protection), Kids Shampoo & Body Wash. Adult products include cosmetics, shaving products and toothpastes. Logona products are available by mail order (p&p free) from Logona UK. Phone 01986 781 782 www.logona.co.uk info@logona.co.uk.

John Masters Organic Hair Products SLS-free shampoos, conditioners and rinses with coconut soap, plant oils and some organic ingredients. Shampoo costs £15.95, Intensive Conditioner costs £22, Citrus Detangler costs £15.95 and Herbal Rinse Clarifier costs £15.40 (all 8 fl.oz.). Available from 21st Century Health*.

Materia Aromatica A large range of essential oils from organically grown and wildcrafted plants are available from this London company, including seven Chakra Blends, which include Free the Spirit (Frankincense, Spikenard and Myrrh) and Inner Reunion (Rose, Palmarosa, Neroli, Lavender and Bergamot) from £4.50 for 5ml and £7.99 for 12ml. There are also bath oils and an organic jojoba oil from £8.10 for 100ml, wild rosehip oil, £12.50 for 100ml, a range of floral waters from £5.50, and Children and Pregnancy starter packs, 10 oils for £45.40, 7 oils for £49.95 respectively. Bulk discounts are available. Order online, by email, post or phone. P&p free within the UK. Phone 020 7207 3461 www.materia-aromatica.com info@materia-aromatica.com.

Natalia Made by Vital Touch, this is a range of mother and baby products using organic and natural ingredients. Products for mother include Prenatal bath soak (200ml for £10.95), Anti-stretch oil (100ml for £14.95), Labour massage oil (60ml for £8.95), New Parent bath soak (200ml for £10.95). Baby products include Baby bath (100ml for £6.45) and Baby massage oil (60ml for £4.95). Kits including several products are available. Natalia can be purchased from Natural Woman* and Ethos. Phone 01803 840670 www.vitaltouch.com.

Natracare Organic cotton tampons (applicator and non-applicator) from Bodywise under the Natracare label are available in leading health stores, by mail order from Bodywise and from other mail order companies. There are 'regular' and 'super' tampons from £1.39 to £2.59 in packs of 10 or 20. Natracare, which had the first UK organic certification for sanitary products, also have sanitary pads and panty shields without plastic top sheets or absorbent gel. Natracare doesn't use rayon or rayon/cotton blends, which are widely used to make tampons, and its cotton is free of chlorine bleaches and additives. Toxic shock syndrome has been associated with ordinary tampons using these ingredients, and dioxins produced during chlorine bleaching are known carcinogens. Natracare has a new range of baby skin products recommended by the Eczema Society, using natural and organic ingredients including Baby Shampoo, Baby Bath and Baby Lotion (all £1.99 for 250ml). Phone 01275 371 764 (Bodywise) www.natracare.com sales@natracare.com.

Natural Alchemy Natural Alchemy of Cornwall makes a range of toiletries using organic ingredients. They include shampoos (£2.49) with organic citrus fruit extracts, Orange & Aloe Vera for normal hair, Lime & Evening Primrose for dry hair, Grapefruit & Ylang Ylang for frequent use, and Lemon & Peppermint for greasy hair. They are available from Bio-D*.

Natural Health Remedies A range of French certified organic essential oils (in recycled Italian bottles), massage oils, shampoos and conditioners and more are available from this company by mail order and in some leading healthfood stores. Essential oils cost from £8 for 10ml, and massage oils from £11.75. There are aromatherapy shampoos and conditioners for adults and children from £8.95, organic floral waters (from £9.75 for 100ml) and even organic aromatherapy chocolates (£4.95). A children's organic aromatherapy kit costs £45.95, and organic moisturising creams cost £9.65. They also sell suncreams. P&p costs £3.50 for orders up to £30, free over £30. Phone 0845 310 8066 www.nhr.kz organic@nhr.kz.

*Neal's Yard Remedies** The little blue bottle range from this long established company, which has over twenty stores and a mail order service, includes Organic Almond Oil (50ml for £5.15), Lavender and Rose Floral Waters with atomiser (£8.70 for 50ml), and a baby range with Baby Barrier (from £5.60 for

For further details about companies marked with an asterisk (*) including how to order, postage & packing charges etc, see the Leading Mail Order Suppliers section on pages 164–167.

40g), Baby Bath (from £5.60 for 100ml), Baby Massage Oil (from £5.35 for 50ml) and Baby Powder with corn starch (£3.95 for 75g). A gift Baby Box has 5 items for £19.75, and a large range of 24 organic essential oils cost from £3.75 for 10ml. Neal's Yard Remedies also sells Australian bushflower remedies, homeopathic tinctures and bulk dried herbs. All ingredients are organic wherever possible. Some products are available from leading independent health stores.

Organic Botanics Certified organic herbs, flowers, essential oils and other organic plant-based ingredients are used in two creamy moisturisers available from this East Sussex company. There are Moisturising Nutritive and Moisturising Nutritive Extra Rich with natural antioxidants, in 60ml jars for £12.45. Ingredients include organic camomile extract, organic cold-pressed jojoba, organic cold-pressed aloe vera, cold-pressed almond oil, and organic essential oil of ylang ylang and lavender. Organic Botanics can be found in leading organic stores, or are available (p&p free) direct from the company itself. Phone 01273 773 182 www.organicbotanics.com.

Simply Gentle The only organic cotton pleats, pads and buds manufactured in the UK, Simply Gentle products are available at most health food stores. Pleats (100g) cost £1.95, 100 pads cost £1.95 and 200 cotton buds cost £2.25. www.simplygentle.com.

Sheer Bliss Natural Soap Handmade natural soap, made with the finest vegetable oils and butters, scented exclusively with essential oils and coloured only with botanicals and natural ingredients. £2.95 per bar, post free. Free samples available: send a stamped address envelope to P. O. Box 541, Folkestone CT20 2WF. Phone 01303 211085 www.Sheer-Bliss.co.uk.

Simply Soaps Organic herbs are among the ingredients used in this range of incredibly delicious-smelling natural soaps available by mail order, from 250 UK stores (list on website) or at Cambridge and Norwich markets. They include Calendula and Lavender, Geranium and Rosepetals, and Seaweed Soap. Prices are £1.90 to £2.25 for each bar. P&p is £2 for up to 4 bars, £4 for 5 or more bars. Phone 01603 720 869 www.simplysoaps.com info@simplysoaps.com.

Soapnut An organically cultivated multi-purpose body washing product in powder form, which has been used for thousands of years in India. A 100gm pot including p&p costs £5.45 from Global Eco. It is also available as bulk loose powder and liquid extract, and can be found in some leading healthfood stores. Phone 01284 700 170 (Global Eco) www.soapnut.com orders@soapnut.com.

Spiezia Organic Care A range of organic baby and adult products, all certified by the Soil Association. Baby cream costs £5.80, baby oil £14.60, and camomile cream costs £7.99. Available from Lollipop, from Harrods, and by mail order. Phone 01326 231600 www.spiezia.co.uk sales@spieziaorganics.com.

Tautropfen This range of luxurious biodynamic skin care products is made by Tautropfen in Germany, the largest Demeter-certified manufacturer of skin and hair products worldwide. The products are highly concentrated and contain no water (most creams contain an average of 70% water). The baby range includes a 30ml Baby Cream (sunflower oil, beeswax and essential oils of camomile and lavender) for £9.40, 50ml Baby Oil (same ingredients without the beeswax) for £7.60 and a Children's Oil (sunflower and olive oils with sun-infused camomile and lavender and essential oils). Products for adults include Rose Water (£8.15 for 50ml), Sea Buckthorn/Jojoba Cream (£14.90 for 30ml), Tea Tree balm (£9.90 for 15ml) and a unique Wash-Clay that can be used on hair and body (£5.10 for 250g). All products are packaged glass containers with wooden tops. Tautropfen is only available from Greenfibres*.

Tisserand Around fifty organic essential oils are available from leading independent health stores or by mail order. The oils range in price from £4.15 to £8.95. There is also Lavender Skin Conditioning Lotion containing organic lavender essential oil (£4.90 for 125ml), a Tea Tree Lip Balm made with organic tea tree oil, and several more bodycare products using organic oils. P&p costs between £1.50 and £2.50. Phone 01273 325 666 (for info and stockists) www.tisserand.com info@tisserand.com.

Urtekram This Danish range of shampoos (Seaweed, Nettle, Tea Tree, Aloe Vera etc for £4.25 and up), conditioners, toothpastes, and other toiletries, including children's lotions and soaps, uses organic, wild and natural ingredients without preservatives. The Urtekram factory uses only power generated by wind and hay. Available in leading healthfood shops. For product range information contact importers Whole Earth. Phone 020 7633 5900 (Whole Earth) www.urtekram.dk (Danish site).

Weleda An extensive UK range of biodynamic and organic and natural baby, child and adult toiletries and herbal/homeopathic preparation. Weleda was founded in 1921 in Switzerland and in 1925 in the UK. Weleda baby care products include Calendula Baby Moisturising Cream (£4.50), Baby Lotion (£4.50), Nappy Change Cream (£4.50), Baby Oil (£5.40), and Baby Soap (£3.70)—using organically grown calendula. Herbs are grown in Weleda's gardens in Derbyshire. Available in leading healthfood shops, from mail order companies, or contact Weleda direct for mail order. P&p is £2.95 (free over £25) and delivery is by first class post. Phone 0115 944 8200 www.weleda.co.uk info@weleda.co.uk.

* * *

Lice Treatments There are several non-toxic, natural lice treatments available from shops and mail order. Quassia chips are sustainably harvested from trees in South America and have natural anti-lice properties. Biz Niz makes a range of preventative treatments including tea tree and quassia shampoo (250ml for £7.50) and conditioner (100ml for £7.95). Biz Niz is available from Natural Woman*. Delacet Scalp & Hair Cleanser contains larkspur and is available from Natural Collection*, 110ml for £7.95.

Non-toxic Hair Colouring There are several brands of natural hair colourants now available. Some are certainly more natural than others, so read the ingredients! Herbal Diffusion Hair Colouring makes Herbatint: Permanent Colour for £8.95 and Vegetal-semi-permanent for £9.95. Nature's Dream makes hair colouring for £7.49. Both brands are available from Revital* and some health food stores. Natural Collection* sells Naturtint for £6.50 a box.

Washable sanitary pads According to the Carrying Kind, disposable menstrual pads and tampons take up a larger volume in landfills than disposable nappies. They also use many trees, release chemicals into oceans and rivers, and expose many women to dioxins in a very sensitive part of their body. There are a number of washable cloth sanitary pads on the market now, some of which are organic. Comfort Zone Sanitary pads are an all-in-one design washable sanitary pads in three flow sizes, starting from £5 and available from Cuddlebabes*. A 12-piece set of washable sanitary pads from Yummies (see Nappies section) costs £16.99. Imse Vimse 'Damindor' washable pads are made from a cotton/polyester blend and have outer envelopes with pads to put inside for greater or lesser absorbency. One pack containing a selection of mini, regular and maxi outers plus inner pads costs £16.99 from Natural Woman*, Twinkle, Twinkle* and The Nice Nappy Company*. Lunapads, made in Canada, come in both conventional and organic cotton options. System includes liner that fit into outers with wings, which snap around underwear. Organic starter kit, which includes 2 regular pads, 2 mini pads and 2 extra liners, costs £29.00 from Twinkle, Twinkle* (also available from Natural Woman*). Eco-babes* sell Ecofem sanitary pads for £17.50 for a pack of three. Other more environmentally friendly options include Luna Sponge natural absorbent sea sponges, to be used as a washable reusable tampon (available from health food stores and direct from Luna Sponges: phone 01458 834787 sweetfa@purpleturtle.co.uk), and The Keeper—a reusable rubber cup that sits inside and collects menstrual flow (available from health food stores and The Nice Nappy Company* for £30).

Organic food supplements

Many leading health experts believe that naturally sourced food and plant-based food supplements are more beneficial and more easily absorbed by the body than separate vitamin and mineral supplements. Some of the supplements listed can be used as a boost during times of stress, increased work, strenuous activity or other demands on the body, such as travel, pregnancy and breastfeeding (seek professional advice first). Others, such as essential fatty acid oils and milled powders, are considered to be essential for health, and for the proper development of the unborn baby during critical pre-conception months, pregnancy (seek professional advice before taking any supplements) and breastfeeding; and for the infant itself from weaning, although you can alternatively buy the seeds yourself and grind these daily on to your food (see Organic Food, Health and Nutrition Tips and The Good Oils in Part One; and *Fats that Heal, Fats that Kill* by Udo Eramus is a useful book on essential fats, available from Savant*). There are now many such quality supplements on the market, including many 'superfood' formulas containing ingredients claimed to significantly improve adults' and children's health, such as spirulina, wheatgrass (which can be grown and juiced instead), chlorella, Klamath Lake Blue Green

Algae, kelp and other nutrient-rich seaweeds. Ask your retailer or the manufacturer about suitability of the supplement for adults, children and babies, and the recommended doses. Most need to be refrigerated after opening, and some must be refrigerated even while unopened. Always seek professional advice before buying and trying any supplements if trying to conceive, already pregnant or breastfeeding, or planning to use any on your weaning infant. If you're following the Foresight programme, supplementation is built into this. Many supplements which can have detoxifying or cleansing effects may not be suitable for use during pregnancy unless taken for some time prior to conception. And general advice is to always introduce a supplement gradually, to allow the body to get used to it. Buying good supplements may mean paying a little more, but there are no artificial additives, and you're getting a quality product which was grown without artificial chemicals and fertilizers. As food supplements can be quite expensive, it is always advisable to make your daily diet your first priority: this should be as nutritious and wholesome as possible before you spend money on supplements, which should not take the place of fresh, whole organic foods if they are available.

Aloe vera The gel and juice of the aloe vera plant is said to be rich in many useful nutrients, has healing abilities and can be a digestive and dietary aid. There are several organic aloe vera products now on the market. Look for juices (which can be taken with fruit juices), tablets and gel skin care from Optima Healthcare, Pro Ma Optimum Health, ESI Laboratories and Higher Nature, and other leading brands. Lifestream Biogenic Aloe Vera Vegicaps contain freeze dried organically grown aloe vera juice from New Zealand. It claims to help with IBS, digestive problems, arthritic pain, and the health and vitality of the skin. It costs £11.50 for 60 capsules from healthfood shops or by mail order from the Xynergy Health Products*.

Hawthorn & Artichoke Formula A liquid herbal food supplement from Green People* (their products are all recommended by Leslie Kenton): a blend of twelve herbal juices, four tinctures and live enzymes and antioxidants to boost your immune system and help absorption of nutrients. A 150ml bottle costs £19.95.

Hemp Oil Hemp is considered to be one of the 'complete foods' and a good source of all the omega essential fatty acids. Organic hemp is now becoming available in this country. Biona Organic cold-pressed hemp oil costs about £8 from independent healthfood stores for a 250ml bottle. Virgin Hemp seed oil is available from Higher Nature (01435 882 880) for £13.90 for 207ml. Mother Hemp is the only commercial hemp grower in the UK, and makes a range of organic hemp edible products including ice cream, pasta and pesto. They also produce raw seeds, toasted seeds, cold-pressed oil and oil in capsules. Mother Hemp products are available at most health food stores or contact direct on 01323 811 909 www.motherhemp.com.

Klamath Blue Green Algae This organic food supplement is claimed to be one of the most nutrient dense wholefood supplements, based on the algae harvested from Klamath Lake in Oregon and with plant source enzymes. Prices start at

£16.90 for 60 vegicaps or tablets, £13.95 for 30g of powder. Klamath Blue Green algae tinctures include Calm-Kid formula (£16.50 for 60ml). It is available in leading healthfood stores, or by mail order from The Really Healthy Company. P&p is £1.50 for orders under £40 and £3 for orders less than £70. Phone 0208 480 1000 www.healthy.co.uk mail@healthy.co.uk.

Linseeds/Flaxseeds Linseeds come from the flax plant (same species as linen)and are a good source of omega fatty acids. Most independent health food stores and mail order supplement companies carry one or more brands of linseeds or flaxseeds. Seeds come whole or milled and may need to be refrigerated. Look for Omega Nutrition Cold Milled Organic Flaxseeds in a vacuum-packed resealable bag. Northern Edge offers Canadian organic milled flaxseed also in vacuum-packed resealable bags for £9.99. Phone 0208 636 9980 www.northernedge.org.uk info@northernedge.org.uk.

Linseed/Flaxseed Oil Linseed or Flax seed oil is the oil that has been cold-pressed from the seeds. It offers a more concentrated source of Omega fatty acids and is good on salads or mixed in with any food (not hot). Widely available in health food stores and some mail order supplement companies. All oils should be refrigerated before and after opening and have a limited shelf life after opening. Biona makes an organic linseed oil (imported by Windmill: 0207 924 2300). US company Omega Nutrition makes an organic flaxseed oil in a chilled 355ml special dark bottle for around £12. Also available in capsules—120 for £14 or 250 for £28. Spectrum Essentials offers organic flax and high-lignan organic flax oils.

Dr Gillian McKeith's Living Food Energy This milled 'sprouted superfood' supplement contains organically grown sprouted millet and quinoa, essential fatty acids from flax, sunflower seeds, nori seaweed, wild blue green algae, herbs and grasses such as alfalfa and barley, and herbs such as aloe vera. Available from leading independent healthfood stores from £19.95. Wholefood bars are also available for £1.65.

Optima Healthcare Manufacturers and distributors of health products including Aloe Vera juice, capsules and tablets. Available in many health food stores. Contact Optima direct for stockist information. Phone 029 2038 8422 www.optimah.com admin@optimah.com.

Other Essential Fatty Acid Oils US company Omega Nutrition makes Essential Balance, an excellent essential fatty acid supplement/organic oil blend to restore health with balanced essential fatty acids in special dark bottles. A blend of unrefined flaxseed, sunflower seed, sesame, borage and pumpkin oils, fresh-pressed using the Omegaflo process. Recommended by Patrick Holford. Also available in dark carob capsules. Omega Nutrition also makes a junior version of Essential Balance, with natural butterscotch flavouring. Costs £8.90 for 237ml. Omega Nutrition products are available at many health food stores and through Higher Nature (01435 882 880). Spectrum Essentials also offers an organic oil blend: organic Essential Max.

ProGreens With advanced probiotic formula, this supplement has been a favourite with nutritionists for some time. It is dairy-free, and contains green organic gluten-free grasses (including wheatgrass juice), blue-green and sea algaes, herbs and 'superfoods' such as wheat sprout powder, beet juice powder, royal jelly and bee pollen. It's available in capsules (180 for £26.10) or powder (30 day supply for £40.60) from The Nutri Centre*.

Pure Synergy Called The Ultimate Superfood Formula, it is a blend of over 60 organic and wildcrafted superfoods from the US including algaes, green juices, freeze-dried herbs, therapeutic mushrooms and much more. Created by Mitchell May after fifteen years' research, it costs £34.50 for 142g or £67.75 for 350g. Available from leading independent healthfood stores and Xynergy Health Products*.

Seagreens Organic seaweed harvested off Norway is available in vegetarian capsules. Seagreens seaweed products are certified by the Soil Association and harvested sustainably, and the seaweed is processed immediately after harvest, on the dock. Seaweed has all the basic nutrients for life in extremely absorbable form, including daily iodine requirement (especially important for vegans). Wild Seaweed Food Capsules are available in healthfood stores and some mail order companies: 60 capsules cost £12.75 (two a day) Up to three of the Wild Seaweed capsules can be taken by adults daily, and in addition to pre-natal or other vitamins & minerals, and the contents can even be sprinkled on to infant's weaning foods from six to eight months. For nutritional or product information and to order (p&p included in price) phone Oceans of Goodness. Other Seagreens products include Seaweed Culinary Ingredient for sprinkling on food or using as a multi-purpose food ingredient and Seaweed Table Condiment which produces a fresh ground version, ideal as a salt replacement. Phone 01444 400403 (for info) 01293 514958 (to order).

Spirulina Certified organic spirulina powder from MicrOrganics in 90g powder costs £14.95, or 100 tablets cost £9.95. Available at leading healthfood stores. Other brands of spirulina products include Earthrise and LifeStream from Xynergy Health Products*.

Superfood Imported from the US, this formula by Dr Richard Schulze is reported to be good for pre-conception, pregnancy, lactation, and for children. It contains Spirulina Blue Green Algae, Chlorella, Purple Dulse Seaweed, Beetroot, Spinach Leaf, Rosehips, Orange and Lemon Peel, and Non-Active Yeast, and is said to provide high protein, all necessary vitamins, minerals, fatty acids and more. A 396g tub which costs £29.50 lasts one month at the recommended dose for an adult, which can be taken with fresh lemon juice and other fruit or vegetable juices. Available by mail order from Herbs Hands Healing: up to £4 postage & packing, orders over £50 free, first class post an extra £1. Phone 01379 608 007 or 0845 345 3727 www.herbshandshealing.co.uk.

True Food Super Nutrition Plus High absorption A-Z multivitamin and mineral complex based on nutrient-rich seaweed and plant bases from Higher Nature. Also available in some leading healthfood stores from £6.60. Phone 01435 884668 www.higher-nature.co.uk info@higher-nature.co.uk.

Udo's Choice* Ultimate Oil Blend: refrigerated organic oils are pressed in a low heat, low light and oxygen-free environment and bottled to provide an excellent source of essential fatty acids. Non-organic ingredients are evening primrose oil, lecithin (GM-free), rice bran, oat bran and germ oils. Udo's is highly recommended by leading experts including Leslie Kenton and Daphne Lambert, and is available in two sizes from £8.95 (for 250 ml) in leading healthfood stores. Keep refrigerated and do not heat. Udo's Choice also makes a Digestive Enzyme Blend. Both products available from health food stores and Savant*.

Viridian This ethical UK supplement manufacturer gives 50% of its profits to charity every year (see list of charitable recipients on website), and they take back their own bottles for recycling. Customers receive 25p for each empty bottle they bring back to the store. High-quality supplements come in vegetarian capsules and include Ester-C (30 capsules for £5.05) and High Five multivitamin and mineral formula (30 capsules for £6.80). Viridian also sells B vitamins, minerals, amino acids, herb extracts, digestive aids and probiotics. Their Organic flax seed oil costs £7.35 for 200ml and their hemp oil costs £8.75 for 200ml. Viridian products are available from health food stores and mail order supplement companies. Contact Viridian directly for product and stockist information. Phone 01327 878050 www.viridian-nutrition.com info@viridian-nutrition.com.

Vita Synergy An all-natural multivitamin containing botanical extracts, wildcrafted and organic flowers and spices and more. Created by Mitchell May who developed Pure Synergy, there are Vita Synergy tablets for Women, and Vita Synergy for Men. Each contains 180 tablets for £54.95. Available from the Xynergy Health Products*.

Wheatgrass Four Seasons Health offers a range of US wheatgrass powders and tablets from leading healthfood stores: 500mg tablets come in bottles of 75 for £6.50, or 150 for £11.75. Whole leaf powder is in bottles of 50g for £9.50, or 100g for £16.50, and juice powder (very concentrated) is available in 50g or 100g bottles for £18.75 and £34.20 respectively. Now Foods offers organic wheatgrass tablets available in leading healthfood stores. Sweet Wheat supplies freeze dried organic wheatgrass juice powder from the US and is available from Xynergy Health Products*. It's ideal for detoxifying, energizing, nourishing and boosting the immune system (ideal during pre-conception), and can be useful for treating eczema and burns because of its high chlorophyll content. It dissolves easily in liquid, and can be given to infants in small daily amounts from weaning. Fresh organic wheatgrass juice is available from many juice bars including Fresh & Wild stores. Or juice your own wheatgrass if you can— starter trays of wheatgrass are available from The Green Seed Company on 0208 789 7897 or 07866 368588. (See Organic Food, Health & Nutrition Tips in Part One, and Juices in Part Two.)

Other high quality supplements

Always seek professional advice before taking supplements. Many nutritionists believe that supplementing with correct doses of quality vitamins and minerals is essential during pre-conception, pregnancy and breastfeeding: they are required in set amounts by the baby, we can't guarantee that we're getting these nutrients or amounts in our food, and exposure to pollution and stress will often deplete the levels in our bodies. Some also believe that babies and children really need these supplements to aid their optimum growth and development. There are other nutritionists and health experts who believe that these 'synthetic' supplements are unnecessary if a balanced organic wholefood diet is being eaten, and that they are not as readily absorbed by the body as food is. These 'opponents' also do not recommend them for young children or babies, as they believe their systems aren't able to cope with them (see Organic Food, Health & Nutrition Tips in Part One).

If you do decide to supplement, ensure that you're taking high quality tablets, capsules, powder or liquids, which are giving you or your baby or child the nutrients you require. Most high quality vitamin and mineral supplements now come in vegetarian capsules or tablets without gelatin—check the label. Vegicaps made from cellulose can be pulled apart so you can use the ingredients as a powder. Powdered supplements, which can be mixed into foods, are available from brands such as Biocare. Several of the best products on the market are listed.

Biocare Offers Ante-natal Complex—a formulated multivitamin and mineral supplement for expectant and lactating mothers from Biocare (practitioner standard range). New in cellulose capsule form with 60 capsules for £10.30. Vitasorb Multivitamin, an easily absorbed child's formula comes in liquid form for £8.25. Also, Vitaforte Multivitamins and Minerals, a child's formula in powder in a base of freeze-dried banana (must be refrigerated) costs £11.80. Available in leading healthfood stores or by mail order. P&p costs £1.95 on orders under £15. Phone 0121 433 3727 (general sales), 0121 433 8710 (customer care) www.biocare.co.uk.

Foresight vitamins and minerals Foresight's specially formulated supplements are recommended for pre-conception, pregnancy and lactation. Foresight vitamins and minerals (each in 90 tablet packs for £4.76, less for members) are sold separately and do not contain iron—there is a separate iron formula. Other supplements include Foresight Zinc, and Vitamin C plus Garlic. They are available from the Foresight Resource Centre. Phone 01483 427839 www.foresight-preconception.org.uk/home-page.html.

For further details about companies marked with an asterisk (*) including how to order, postage & packing charges etc, see the Leading Mail Order Suppliers section on pages 164–167.

Health Plus Pregnancy Pack is a multivitamin and mineral supplement and costs £12.25 for a month's supply. Supermouse is a children's multivitamin formula in 90 or 200 tablet size for £5.95 or £11.95. Both products are recommended by Patrick Holford, Hazel Courtney and the Institute of Optimum Nutrition. Available by mail order from Health Plus. P&p free on orders over £5; 50p under £5. Phone 01323 737 374.

Higher Nature Makes Dynochews, a chewable quality multivitamin and mineral for children, which makes taking necessary supplements more fun. According to HN, their chews contain several times the essential minerals of many children's formulas, and are flavoured with real fruits and honey. Available in 30 or 90 vegetarian tablet packs. Advanced Nutrition Formula is nutritionist Patrick Holford's ultimate multivitamin and mineral formula with citrate minerals for easy absorption (30 tablets for £5.10). Higher Nature donates a percentage of its profits to charities. Phone 01435 884668 www.higher-nature.co.uk info@higher-nature.co.uk.

Solgar Pre-natal Nutrients are part of the large Solgar range. Available in bottles of 60 and 120 tablets for £6.45 or £11.45. VM2000 is a powerful multivitamin for £8.79 for 30 tabs and Vita-Kids wafers are a children's multivitamin for £5.95 for a bottle of 50 tabs. Available from health food stores and the Nutri-Centre. Contact Solgar for product and stockist information. Phone 01442 890 355 www.solgar.com solgarinfo@solgar.com.

Natural remedies

Organic herbal and organic and natural homeopathic remedies are now readily available for many ailments in place of common over-the-counter drugs, but always seek professional advice before using them. Homeopathic remedies will contain either lactose or sugar; herbal tinctures will be alcohol-based. There are many useful free booklets (and books you can buy) on how to get the most out of these remedies, available from Weleda, Boots, Neal's Yard Remedies, Ainsworths and many other mail order catalogues featured in this guide, or from healthfood stores and bookstores. Books include *Natural Healing for Women* by Romy Fraser and Susan Curtis (£12.99), and *Homeopathy for Mother and Baby* by Miranda Castro (£15.99)—both available from Neal's Yard Remedies*. The Soil Association has set organic standards for 'health and beauty' products, which will cover some of the products listed here.

Ainsworths This leading homeopathic pharmacy, based in London and Caterham, Surrey, has a large range of homeopathics in-store, by mail order, and an over-the-counter range available in pharmacies and healthfood stores. Remedies cost from £4.50 for 120 tablets for the counter range, and from £2.00 for in-store or mail order products. The whole spectrum of homeopathic formulations is available from Ainsworths. There are also first aid kits which can fit into a handbag in a leather wallet (from £29) for childbirth, children, family and travel. There's an emergency kit which can be carried in a pocket or handbag in a leather wallet for £39, ointments, creams, lotions, liniments, and Ainsworths' Bach Flower Remedies from £23.15 each. Contact Ainsworths for information on homeopathy: they can supply a register of homeopaths, and many homeopathic books (including ones on vaccination). P&p for mail order products is between £2 and £3. Phone 0207 935 5330 (London) or 01883 340332 (Surrey).

Australian Bushflower Essences A large range of flower extracts from the Australian bush, available singly (15ml for £7.95) or in combination to help with the ups and downs of life (30ml for £7.95). Available from Neal's Yard Remedies* and some leading healthfood stores.

Bach Rescue Remedy Flower extracts first used by Dr Edward Bach in the 1930s can be used diluted in water or taken straight on the tongue; Rescue Remedy is recommended for labour and any other unsettling situation. It costs from £3.15 for 10ml. Flower Essence Cream is also available for £3.20 for 30g (ideal for dry, chapped skin and nappy rash). Available in leading independent healthfood stores, Neal's Yard* and Boots. All single Bach flower remedies are also available in 10ml for £2.90 from leading stores.

Bioforce A huge (and the leading) imported range of organic herbal tinctures— created by renowned Swiss nutritionist, herbalist and naturopath Alfred Vogel— suitable from six years upwards. Promoted by leading naturopath Jan de Vries, there are phytotherapy tinctures, and also restorative creams made with echinacea, St John's Wort Oil and herbs. Echinaforce drops are an organic echinacea tincture—ideal for boosting the immune system and maintaining resistance to winter's coughs, colds and flu for children and adults. Fifteen to twenty drops are recommended three times a day to build resistance or to fight off a cold or flu in early stages. Add to a small glass of water. A 50 ml bottle or 120 tablets costs £7.49. Bioforce products are found in most independent health stores, or contact Bioforce direct for stockist information. Phone 01294 277344 www.bioforce.co.uk.

Boots Homeopathic Range Boots now has a large range of own brand homeopathic remedies for simple ailments. An easy-to-use guide appears alongside the products on the shelf. See them at your local store.

Digestive Enzymes These can be taken on their own or as part of a food supplement, to aid absorption of nutrients. Most leading supplement brands will be able to provide digestive enzymes.

Hambleden Herbs Organic herbal tinctures made by The Organic Herb Trading Company can be added to ointments or taken in water. Tinctures include Milk Thistle, Ginkgo, Marigold and St John's Wort. A 50ml tincture costs from

For further details about companies marked with an asterisk (*) including how to order, postage & packing charges etc, see the Leading Mail Order Suppliers section on pages 164–167.

£5.90. There are also organic comfrey ointment and comfrey oil, eighteen Healing Herbs Flower Essences (made in Wales following Dr Bach's methods), a Five Flower Remedy, and Flower Essence Cream. Single essences cost from £2.50 each for 10ml, a complete set costs £85, the flower remedy costs from £2.50 for 10ml and a 30g cream costs from £3.20. Available through independent healthfood stores.

Helios Homeopathic Kit This easy-to-use kit for the family costs £24.95 and is available from leading independent healthfood stores. It contains eighteen homeopathic remedies in bottles to treat all family ills. A basic guide to homeopathy is included. The childbirth kit also contains eighteen remedies and costs £25.95 and has useful recommendations for labour pains and post-natal care for the mother and baby, including use of Arnica, Chamomilla, Aconite and Calendula.

Herbs Hands Healing Formulas Herbs Hands Healing uses many of its own organic and wildcrafted herbs to create individual formulas including those, which will ease morning sickness symptoms and help prevent miscarriages. Other formulas can be made to suit individual requirements, and advice is also available. Formulas are available by mail order (orders over £50 post-free). Phone 01379 608007 or 0845 345 3727 www.herbshandshealing.co.uk.

Jan de Vries Jan de Vries Phytotherapy Range offers organic herbal preparations including Aloe Vera complex (to help psoriasis), Black Cohosh, Feverfew, Ginger, Kava-kava and Peppermint Complex (to help irritable bowel syndrome). There's also a range of unique combination flower essences to help with modern health problems. They cost £7.50 each for 50ml and are available from healthfood stores. For stockist enquiries and customer information contact Bioforce.

La Drome Provençale A range of imported organic remedies, including flower essences and Propolis products. Organic Flower Essences are prepared according to Dr Edward Bach's remedies: forty are available at £3.65 each (10ml). La Drome Provençale also makes a first aid cream with organic flower essences for £4.45, and an organic alternative to Vicks Vapor Rub called Fyto-rub: a 45g tub costs £5.25. There are mouth and nasal sprays, and Propolis syrups with herbs such as Echinacea. The remedies are available in leading healthfood stores and also by mail order from importers Windmill (020 7924 2300).

Little Miracles A range of six flower essences designed specifically for babies and children that can be put on the tongue or the skin. Remedies include 'Chill-out' for calming and relaxing, SOS (like rescue remedy), both for £7, and Soft Touch healing gel for irritable skin conditions, which costs £8. Available from 21st Century Health*, Earth Mother or direct from Little Miracles. Phone 020 7431 6153 www.littlemiracles.co.uk.

Nature's Answer A large range of organic and wildcrafted herbal extracts from the US. Full ranges at leading healthfood stores. Excellent products include Nat-Choo for nipping children's colds in the bud (118ml costs £10.95), which can be diluted in smaller amounts for older babies. It contains echinacea, catnip, peppermint, boneset and eyebright in a base of coconut glycerin and citric acid from cassava root. There are also Windy Pops, Kiddy Tum, Immuno Kid and E-Kid-nacea. www.naturesanswer.com (US site).

Neal's Yard Remedies* Neal's Yard offers one of the largest ranges of natural remedies using organic and wildcrafted ingredients, including its own tinctures, homeopathics, ointments, macerated oils etc, and North American Flower Essences, Bach Flower Remedies and Australian Bushflower Essences. Over 100 herbal tinctures made with the company's own organically grown plants and local spring water are available, in 50ml (from £4.60). There are also homeopathic tinctures made with organic herbs (15ml for £4.45) and single homeopathic remedies (50 tablets for £4.50), and a huge selection of dried herbs.

Nelsons This established brand has Teetha, a homeopathic treatment for infant teething pain, containing Chamomilla and lactose, costing £4.20 for a packet of sachets from healthfood stores and Boots.

New Era An established large range of tissue (mineral) salts used homeopathically to treat all manner of ailments, including Coughs, Colds & Chestiness, and Infants Teething Pain (£2.39). There are eighteen combination remedies. Found in most independent healthfood stores.

Olbas Oil A non organic but very useful blend of powerful essential oils for easing congestion and making breathing easier when coughs and colds set in. Some of these oils are not recommended during pregnancy or for very small babies. Use in moderation away from direct contact or inhalation. Available from healthfood stores and pharmacists. www.olbas.com (US site).

Probiotics or beneficial bacteria A course of antibiotics, times of stress and illness or poor diet can deplete our natural gut and bowel bacteria. Probiotics or beneficial bacteria are often recommended to restore these. Brands to look for include BioCare (for adults and babies and children). Phone 0121 433 3727 (general sales), 0121 433 8710 (customer care) www.biocare.co.uk.

Weleda An extensive UK range of biodynamic herbal and homeopathic preparations for all types of ailments is available from Weleda, founded in 1921 in Switzerland and in 1925 in the UK. Organic herbs are grown in Weleda's gardens in Derbyshire. Available from healthfood stores, pharmacies or from Weleda direct. Details of doctors, clinics and therapists specializing in anthroposophic and homeopathic medicine in the UK are available from Weleda. P&p is £2.95 (free over £25) and delivery is by first class post. Phone 0115 944 8222 www.weleda.co.uk info@weleda.co.uk.

For further details about companies marked with an asterisk (*) including how to order, postage & packing charges etc, see the Leading Mail Order Suppliers section on pages 164–167.

HOUSEHOLD PRODUCTS

Water filters

Unfortunately our drinking water is not pure—it's contaminated with many pollutants. These include pesticide and nitrate residues from farming and from use on lawns, golf courses etc, which reach the groundwater and waterways; heavy metals such as lead from old pipework; bacteria; hazardous pollutants leaking from landfill and industrial waste sites into groundwater and waterways; and treatments at water treatment works such as aluminium sulphate, chlorine and fluoride. Bottled water is also largely an unknown quantity, and pollutes the earth with plastic bottle production and packaging waste, and transportation. It's also thought that the plastic can affect water quality. While some purists advocate using distilled water to produce truly pure water, well-filtered water can remove most if not all pollutants. In the US, research has shown that we actually absorb more contaminants through bathing and showering than from drinking polluted water. A child can absorb up to 50% of contaminants from drinking or eating, while through skin exposure and inhalation of steam children absorb nearly 100% of contaminants directly into the bloodstream. Drinking and bathing in pure, well filtered water can help to clear up some health problems and is important as part of any pre-conception or cleansing programme, during pregnancy, breastfeeding and for your baby.

Aquathin Recommended and used by The Institute of Optimum Nutrition, The Hale Clinic and several hospitals. Aquathin, made by The Pure H2O company, offers portable, non-plumbed-in and plumbed-in reverse osmosis and de-ionisation water purification systems to either buy or hire. Systems cost from £399, or can be rented for as little as £3.63 a week—less than the weekly cost of bottled water. Aquathin Pure H2O is becoming the leading brand of water sold in refillable containers throughout the major healthfood stores—if you're unable to buy or hire a filter, consider stocking up during your weekly shop. Aquathin can also deliver water in bottles to your home or office if you live in Greater London, parts of Surrey, Hampshire, Berkshire and Hertfordshire. Phone 01784 221 188.

Brita or similar For families on a tight budget, Brita or similar brand jugs and separate charcoal filters are available in healthfood stores and other retail outlets. These remove some pollutants—probably the cheapest filter option. It's vital to change the filter regularly.

Fresh Water 1000 Fresh Water Filter This filter, highly recommended by leading nutritionists, has a six-stage ceramic filtration process which removes most pesticides, herbicides, oestrogen, fluoride and nitrates, chlorine etc. Available from Savant Distribution* or The Healthy House* for £249.95 (installation is £55 plus VAT). Replacement cartridges cost £29.49. The people who make Fresh Water, and many of the mentioned stockists, can also provide a shower filter for £61.10.

Plumbed-in water filter Mounted under the sink with fountain tap fitted on the sink. Tubes and fittings are supplied. A replaceable cartridge lasts for six months. UK-made, and available from Natural Collection*, the filter costs £199 plus £29.95 for a filter refill.

• Natural Collection* offers a shower filter for £68.50 (£37.50 for replacement cartridge) and The Wholistic Research Company (01438 833100) offers a range of water distillers and a shower filter. The Fresh Network (0870 800 7070) also offers water distillers, as does The Healthy House*.

Juicers

There are many excellent quality juicers on the market. The best are large (sewing machine size), and cost from around £300 to over £400—which can be quite a financial investment, although the better the juicer, the more and better quality juice you will get. Some of the leading juicers can also homogenize, grate, grind, and some can juice wheatgrass. They include the Champion Juicer, Green Star Juicer, Suco Juicer and the Angel Juicer. These, and cheaper but quality models such as the Juice Master, are available from several stockists who can deliver, including Organics Direct*, The Fresh Network (0870 800 7070), Savant*, UK Juicers.com (01904 704705, www.ukjuicers.com) and The Wholistic Research Company (01438 833100).

Ask many questions about your juicer-to-be—what can it do? How efficient is it? Is it easy or difficult to assemble and wash? What are the advantages over other models, and the disadvantages? Are there attachments for the juicer to do other tasks? Get brochures and consider well before deciding.

There are also much cheaper manual juicers for foods such as wheatgrass (for example the Porkert Wheatgrass Juice Extractor for around £30) from several of these stockists.

Sprouters

While it's easy enough to make your own sprouter for sprouting seeds such as alfalfa out of a glass jar, you might like to try a couple of products on the market designed to make your life easier. There are Glass Sprouting Jars with mesh lids and stands so you can simply soak, rinse and stand to drain. They come on their own (some with an individual stand), or as a two or three glass jar kit with a two or three glass drainage rack and tray. Prices start from £8.95 plus p&p. Stockists of sprouters such as the Eschenfelder Glass Sprouter kits or Freshlife Automatic Sprouter include The Wholistic Research Company (01438 833100), The Fresh Network (0870 800 7070) and many health food stores. Wholistic Research Co. also sell organic seeds for sprouting and wheatgrass growing kits.

THE 'GREEN' HOME

British studies have found that the air inside your home can be up to ten times more toxic and polluted than the air outside, even in busy London. There are many hazardous chemicals, substances and materials used in modern homes which contribute to this: kitchen, bathroom, toilet and laundry cleaners and bleaches, hydrocarbon aerosols (hair sprays, air fresheners), polishes, furniture and nail varnishes and paint and nail varnish removers, chemicals and solvents in wallpapers, paints, carpets and carpet treatments, pet treatments, MDF (also used in some dolls' houses), plywood, asbestos, pesticides in timber treatments, and many more. We use many of these products every day without ever considering their effects. Some of them do not even list their ingredients, but may be behind mysterious headaches, rashes and cases of asthma, eczema and other ailments. (A recent UK study, featured in 'More than just a food issue' in Part One, confirmed the incidence of this among pregnant women and babies.) Cleaning products and aerosols carry up to eleven warnings such as 'irritant'; 'do not breathe in spray'; 'inhalation can be harmful or fatal', and 'use in well ventilated areas'. No wonder they also say 'keep away from children'! We breathe in their toxic fumes and absorb them through our skin. Many are also causing harm to the environment and killing wildlife when they escape into waterways, or into the air, damaging the ozone layer and diminishing air quality.

Many chemicals in the home will make a pregnant woman nauseous, and if used, could potentially harm the foetus. Many are dangerous for babies and young children to inhale, and can kill if ingested—over 30,000 young children are reportedly hospitalized each year in the UK as a result of contact with common domestic substances. Sadly, many expectant or new parents rush to paint and decorate the baby's nursery close to its birth date, not realizing that the resulting fumes can harm the foetus or newborn. Strong-smelling chemical residues of regular laundry detergent brands on your newborn's clothes could also irritate his skin and affect his health, perhaps contributing to eczema or asthma.

There are also many hazardous chemicals and materials used in the garden: herbicide-treated seed, weedkillers and pesticides, and timber treatments containing the toxic pesticide lindane (listed as gamma HCH), which is linked to breast cancer, and which several environmental groups are trying to ban.

If you do an environmental 'stock take' of your home and garden products, you may be shocked at how many hidden dangers you find. Use this guide and the excellent specialist books on the market to replace these products and green your home for your baby and the rest of your family.

Thankfully there are now organic and environmentally friendly alternatives available for you to replace common household and garden products and materials—ideally during pre-conception and certainly before your baby is born. Listed are cleaning agents and paints, and seed sources for healthy plants to grow to eat and enjoy. There are also many useful books on how to green your home and garden, or even build a green home from scratch. They include *The New Natural House Book* by David Pearson (Conran Octopus), *Eco-Renovation: the ecological* *home improvement guide* by Edward Harland (Green Books), *H is for EcoHome: an A-Z Guide to a Healthy, Planet-Friendly Household* by Anna Kruger (Gaia Books) and *G is for EcoGarden: an A-Z Guide to an Organically Healthy Garden* by Nigel Dudley and Sue Stickland (Gaia Books). All of these are available from the Centre for Alternative Technology; also see the information sources on Organic Gardening. Other things to consider to green your home (which are not included in the listings) include composting your kitchen and household waste. *Backyard Composting* by John Roulac (Green Books) is an excellent guide to how to do this, or you can buy a Can-O-Worms worm composter from Wiggly Wigglers, which is available from The Green Shop* in Stroud or Organics Direct*. You should also consider using energy-saving lighting and appliances: see catalogues from Natural Collection*, The Green Shop*, the Centre for Alternative Technology* or The Green Building Store (Phone 01484 854898 www.greenbuildingstore.co.uk info@greenbuildingstore.co.uk) For good advice and further information on all aspects of green building get in touch with the following sources:

Association for Environment Conscious Building (AECB) AECB is a not-for-profit trade association promoting green building. They have a large selection of books and can provide advice on all aspects of green building. Phone 01559 370 908 www.aecb.net info@aecb.net.

Construction Resources Britain's first ecological builders merchant and building centre, Construction Resources offers advice, mail order and has a showroom in London (16 Great Guildford Street, London SE1). Green building and household materials available include unfired clay bricks, wool and recycled newspaper insulation, solar systems, water saving devices, rainwater harvesting systems, natural plasters and paints, non-toxic wood flooring, untreated wool carpets. Books also sold. They offer a number of tours, courses and seminars throughout the year. Products are available to order by phone. Phone 020 7450 2211 www.constructionresources.com sales@constructionresources.com.

Precious Earth Products and advice for ecologically sound building Phone 01584 878633 www.preciousearth.co.uk info@preciousearth.co.uk.

You could also consider using solar power and other renewable energy systems: contact campaigning group Solar Century on 020 7803 0100, The Green Shop* or the Centre for Alternative Technology*. The latter two will also be able to give you information on water saving and recycling systems. There are a number of companies now offering renewable energy systems for home use. The most common systems available are solar hot water heating and photovoltaic (or solar electric). The following companies offer several different types of these two systems. Solar Services (01985 218899), Solarsense (01225 874299, solarsense@cableinet.co.uk), Solartwin (01244 403 407, www.solartwin.com), The Organic Energy Company (0845 458 4076, info@organicenergy.co.uk), The Solar Design Company (0151 606 0207, www.solardesign.co.uk).

Cleaning agents

Some common cleaning agents and laundry powders carry the warning 'irritant' on them. Why clean surfaces that your baby is going to come into contact with, or wash clothes which are going to be near his skin (or yours), with an irritant? Pungent artificial fragrances, phosphates, sulphates, bleaches, petrochemicals and many other harmful and artificial ingredients (including genetically engineered enzymes) add to the total chemical load on you and the environment and are simply unnecessary. Just walking down a shop aisle lined with these chemical household products is enough to make you gasp and your eyes water. For your own and your baby's health, avoid them at all costs!

Aqua Ball   An environmentally friendly alternative to detergent and softener. Produces ionized oxygen that increases the pH level of the water, activating water molecules so that they lift dirt away without damaging fabrics, and lasts over 60 washes. Hypo-allergenic and comes in a double pack for £14.95, so each wash costs about 11p. An Aquaball Stain Remover is available for £9.99. Contact importers 21st Century Health*: there is a money-back guarantee. P&p is £3. Phone 020 7935 5440 or 0800 026 0220 (to order) www.aquaball.com info@21stcenturyhealth.co.uk.

*Bio-D**   Made in the UK since 1988, these excellent cleaners are all biodegradable with no phosphates, chlorine bleaches, titanium dioxide, synthetic perfumes, glycerin, etc. The range includes Washing Powder (£5.60 for 2 kilos), Washing Up Liquid (£1.78 for one litre), Toilet Cleaner (£2.50 for one litre) and Multi-Surface Cleanser (£2.49 for one litre). Larger volumes are available. The products can be found in many independent healthfood stores, and can be ordered by mail order from Bio-D Direct* or from The Natural Collection*. (Bio-D also do traditional handmade soaps with essential oils.) Recommended by the Breakspear Hospital, specialists in the treatment of allergies and related illnesses. Bio-D products are also available from The Green Shop*, and Concentrated Washing Powder is available from Eco-Babes*.

Dr Bronner's Liquid Soap   Imported from the US, there are eighteen uses listed for this super-mild cleansing soap, which comes in fourteen different guises including a baby super-mild version, almond oil, lavender and eucalyptus oil versions. Ingredients include potassium solution, coconut, jojoba and olive oils. Dr Bronner's products have been popular in the US since the 1960s. Prices start from £2.50 for liquid soap and £2.88 for soap bars. Order from importers 21st Century Health* on 0208 420 6474 P&p is £3. Phone 020 7935 5440 or 0800 026 0220 (to order) www.aquaball.com info@21stcenturyhealth.co.uk.

For further details about companies marked with an asterisk (*) including how to order, postage & packing charges etc, see the Leading Mail Order Suppliers section on pages 164–167.

Citra-Solv This is a multi-purpose cleaner and de-greaser imported from the US which contains citrus oils and can be used to clean a wide variety of surfaces, either diluted in water or used full strength—even bike chains, ovens and carpets. Prices start from £1.50 for Citra-Solv Concentrated Multi-purpose Cleaner (2oz) up to £11.40 for a 32oz bottle. Order from importers 21st Century Health*. P&p is £3. Phone 020 7935 5440 or 0800 026 0220 (to order) www.aquaball.com info@21stcenturyhealth.co.uk.

Clean House, Clean Planet Kit Available from 21st Century Health*, this kit includes bottles and Dr. Bronner's pep soap to make household cleaning recipes in the book *Clean House, Clean Planet*. The book is also available as well as oils used for other recipes. Basic kit costs £17.95.

Clear Spring Range of washing products including Washing-up liquid (one litre for £1.76), Laundry liquid (one litre for £2.90), Dishwasher Gel (£3.90). Available from health food shops, The Green Shop* and Natural Collection*. Phone 0161 764 2555.

Ecoballs Three chemical-free laundry balls which replace all soaps and detergents will last for 750 washes (claimed to only cost 3p a wash compared to about 30p for regular detergents). The balls, endorsed by Action Against Allergy, produce ionized oxygen which activates the water molecules so that they penetrate clothing fibres and lift dirt away. The balls are hypo-allergenic, antibacterial and fragrant-free, leave no residues in fabrics and cost £34.99. They are said to save water and electricity, as you apparently don't need a second rinse cycle to wash out detergent. You can now get pellet refills for them. Order from the Centre for Alternative Technology* or EcoZone. Phone 020 8662 7200 www.ecozone.co.uk info@ecozone.co.uk.

Eco Lino An imported range of biodegradable, phosphate and chlorine-free cleaning range containing vegetable ingredients. Products include laundry powder, fabric conditioner, wool wash, and liquid laundry soap. One litre of washing up liquid costs £1.75, and Multi-surface cleaner costs £2.19. Contact importers Windmill (0208 924 2300) for stockists. Available in some independent healthfood stores.

Ecover Natural botanical and mineral ingredients gentle on the hands and the environment are used in this cleaning brand, free of phosphates and petrochemical detergents, which is well established and respected in the UK. Ecover has solar powered, turf covered roofs on its company buildings, and runs promotions with the Meadowlands Trust to preserve ancient meadows. Made in Belgium, the range includes Washing Up Liquid (£1.95), Concentrated Washing Powder (from £5.25), Concentrated Fabric Conditioner (from £2.49), Dishwater Tablets (£4.38 for a pack of 25) and Multi-surface Cleaner (£1.35). Available from major supermarkets, most independent healthfood stores, some products from Natural Collection*, The Green Shop* or contact Ecover Direct. Phone 01635 528 240.

ECO-wash Three washing disks for £15. Available online or by phone from Yanix. P&p is £1.50. Phone 01388 604958 www.yanix.co.uk/eco-wash/.

Microfibre cleaning cloths All you need is water to clean effectively with these amazing cloths. From many health food stores and mail order companies including Natural Collection* (£14.95 for a set of 4 cloths: floor cleaner, sponge and two multi-purpose cloths) and The Healthy House* ('Clever Clean' all-purpose cloth costs £5.50, mop system costs £36). A full range is also available from manufacturer EnviroProducts Ltd. including a general purpose cloth (in five colours) for £4.45, glass/polishing cloth from £4.95, mop set for £24.95. CD/optical cloth, car cleaning pack, window cleaning pack and large sizes also available. Phone 01892 752199 www.enviroproducts.co.uk.

Nappy Fresh Environmentally friendly, biodegradable and very effective (probably the best on the market) nappy sanitizer and stain remover made in the UK, from Bio-D*. Allergy tested by 'allergy aunt' Claire Fretwell. Antibacterial and germicidal, it is hypoallergenic, enzyme, perfume and chlorine bleach-free. Costs £2.25 per 550g including p&p from Bio-D*. Available in some independent healthfood stores, from Natural Collection* and from Eco-babes*.

Soapnut An organically cultivated multi-purpose cleaning powder, used for thousands of years in India. It is said to remove chemical residues from conventional foods and cottons. A 100gm pot incl p&p is £5.45, and there are also half kilo and kilo bags available. Also found in some leading healthfood stores. Phone 01284 700 170 (Global Eco) www.soapnut.com orders@soapnut.com.

Turbo Plus Ceramic Laundry Disc Magnets, bioceramics and copper microfibre are reputed to produce the cleanest possible, most hygienic wash without the use of soap powders, ionizing the water, reducing water surface tension and lifting out dirt particles from the fabric. It lasts for over 700 washes (3p a wash) and costs £45. Contact distributors Savant* (maximum £5 delivery charge).

Veggi-wash This cleaner removes pesticides from the surface of fresh fruit and vegetables, and can also be used on organic produce to remove any surface grime; it rinses away completely. Surfactants from natural fruit and vegetable extracts are used. It costs £2.99 for a 550ml container from healthfood shops, or can be bought by mail order from Food Safe Ltd with an added £1 for p&p. Phone 01788 510 415.

Washing Disc This reduces the use of washing powder, colour fading and reactions to detergents. The disc contains activated ceramics and magnets which treat the water, allowing you to wash at lower temperatures and reduce the rinse cycle. Lasts for over 500 washes and is 10cm in diameter. Made in the US. Costs £45 from The Natural Collection* and there's also a stain remover (papaya enzyme cleaner).

• You can also make your own cleaners with ingredients such as borax, bicarbonate of soda, tea tree oil or eucalyptus oil and vinegar. See *Clean House, Clean Planet* on page 175.

Paints

No more getting high on paint fumes: avoid solvents and other petrochemicals which can be harmful to you and your baby's health, and instead use organic and natural solvent-free brands which look great, won't cause health problems, or damage the environment. As well as paints, look for non-toxic varnishes, paint strippers, etc.

Auro Natural Organic Paints Made from 100% natural oils and resins with some organic ingredients, Auro offers a range of non-toxic paints, waxes and varnishes. Mail order only. P&p is either £5 or £8 depending on quantity. Also available from The Green Shop*. Phone 01799 543077 www.auroorganic.co.uk sales@auroorganic.co.uk.

ECOS Organic Paints A solvent-free range with white, natural white and eighty-four colours. They also have a special paint for sealing in MDF to absorb formaldehyde and other VOCS, another for removing pollutants from indoor air and still another for shielding against electromagnetic radiation coming in the walls of the house. Nationwide delivery is available, p&p is £7 plus VAT. Sample pots are available. Also sold by The Healthy House*. Phone 01524 852371 www.ecospaints.com mail@ecospaints.com.

Ecotec paints This range of natural paints is made by Natural Building Technologies in the UK. They have a range of pre-mixed and white emulsions and glosses as well as powder and liquid colour pigments. Ecotec paints are available from The Green Shop*. Natural Building Technologies also manufactures wood oils, waxes and treatments and natural building materials such as earth plaster, clay bricks and reed board. Phone 01844 338338 www.natural-building.co.uk info@natural-building.co.uk.

Green Paints Made in the UK for The Green Shop*, these paints, though not entirely natural contain no petroleum solvents, vinyl or chlorinated polymers. This paint range won the *Which* magazine 2002 'Best on Test' award.

La Tienda Earth and Mineral Pigments of Cornwall sell these Mediterranean earth and mineral pigments which can be mixed with natural or conventional paints. Also available from The Green Shop* or phone La Tienda. Phone 01736 360 788.

Livos Natural Paints & Timber Care Non-toxic, pesticide-free, linseed oil-based formulas imported from Germany. Phone 01952 883 288.

Liz Induni Traditional Paints Natural and pigmented limewash by mail order. Phone 01929 423 776 induni@tesco.net.

Nutshell Natural Paints Non-toxic milk-based powder paints, herb and resin oils, waxes for floors and furniture. P&p from £1 for a sachet by post to £7.50 for a larger order by courier. Orders over £200 are delivered free. Phone 01364 73801 www.nutshellpaints.com info@nutshell.paints.com.

OS Color This solvent-free paint range from German company OSMO uses natural vegetable and plant oils, mineral pigments, no biocides or preservatives and is safe for use on children's toys. Winner of Germany's Blue Angel Award, OSMO also offers wood stains and waxes, including an organic wood stain from the woad plant. Available from The Green Shop*. Contact OSMO for stockist and product information. Phone 01296 481220 http://194.143.182.100/osmo/ info@osmouk.com.

Other Household Products

Biodegradable plastic bags These bags are now available from stores in refuse-size and smaller sizes. They degrade naturally over time into CO_2 and water. Both Tuffy and Ecover brands are available from shops.

Candles Most candles are made from waxes derived from petroleum, and let off fumes while burning. Candles from beeswax and vegetable waxes are natural, non-toxic alternatives. These can be found in many health food shops and through several mail order companies including Ecozone and The Green Shop*. Brands include aromatherapy range Nature's Candles from Blake & Fox in Bristol, Circaroma, Neal's Yard and Moorland candles.

Fired Earth A large range of natural floor coverings such as coir, jute and seagrass, which can reduce dust and allergies and are extremely durable, can all be found in the catalogue from Fired Earth. P&p is based on weight. Phone 01295 814300 www.firedearth.com enquiries@firedearth.com.

One Village Natural and fair trade goods for the home including cotton, jute and sisal rugs. Order online or visit one of their three outlets. P&p costs £3.50 for orders under £100. Phone 0845 4584 7070 www.onevillage.co.uk progress@onevillage.org.

Seeds, plants & gardening supplies

Ferme de Saint Marthe Sells rare and forgotten varieties of seeds. Phone 01932 266630 chaseorg@aol.com.

Fertile Fibre Don't be tempted to use peat from endangered peat bogs. If you can't make your own compost, try organic coir compost from coconut trees—coir blocks which turn into compost with added water. Also sell many certified organic gardening products. Phone 01584 781575 www.fertilefibre.co.uk sales@fertilefibre.fsnet.co.uk.

The Green Seed Company This company provides trays of wheatgrass (which last about a week with watering) throughout the UK by overnight courier to households and juice bars. Phone 020 8789 7897 info@green-seed.freeserve.co.uk.

Jekka's Herb Farm Jekka McVicar, a leading organic herb grower, supplies more that 450 species of herb plants and112 varieties of herb and spice seeds to grow. Phone 01454 418878 www.jekkasherbfarm.com farm@jekkasherbfarm.com.

The Organic Gardening Catalogue The official catalogue of the Henry Doubleday Research Association, this catalogue (both in paper and online) is packed with over 400 varieties of organic seeds, books and even insects, which prey on pests. Phone 024 7630 3517 www.hdra.org.uk enquiry@hdra.org.uk.

Pinetum Products Direct They have two catalogues: a Genuine Organic Seed Catalogue which includes vegetables and herbs, and a Mail Order Catalogue, which features all sorts of organic and 'conservation' fertilizers, books, wormeries, vitamin supplements and more. Phone 01452 750 402 pinetum1@aol.com.

Stormy Hall Carries Demeter-certified biodynamic seeds, many of which are grown in the UK. Phone 01287 661369.

Suffolk Herbs Their catalogue has a huge range of organic seeds, herbs, wild flowers, vegetables etc. Phone 01376 572 456 www.suffolkherbs.com sales@suffolkherbs.com.

Tamar Organics Have a large catalogue selling all kinds of organic seeds and garden supplies. Phone 01882 834887 or 01822 834284 (orderline) tamarorganics@compuserve.com.

Terre De Semences Organic vegetable and flower seeds, based in Canterbury. www.terredesemences.com.

Edwin Tucker & Sons Seed specialists for over 170 years, with one of the largest certified organic seed varieties. Catalogue comes out every October. Phone 01364 652233.

Wiggly Wigglers Turn your household waste and kitchen scraps into useful compost with a handy wormery using native British worms. Can-O-Worms are also available from The Green Shop*, and www.getethical.com. Phone 0800 216990 (wormline) www.wigglywigglers.co.uk wiggly@wigglywigglers.co.uk.

For further details about companies marked with an asterisk (*) including how to order, postage & packing charges etc, see the Leading Mail Order Suppliers section on pages 164–167.

LEADING STORES

The following are a few of the many leading stores around the country selling organic food, natural remedies and other organic and environmentally friendly products featured in this guide. For a full list consult *The Organic Directory*.

Ainsworths This leading homeopathic pharmacy in London and Caterham (Surrey) is a useful port of call for information on homeopathy, a register of homeopaths and products in-store and by mail order. There's a full range of useful homeopathic products and books. P&p is between £2 and £3. Phone 0207 935 5330 (London) or 01883 340332 (Surrey) www.ainsworths.com.

Cook's Delight Winner of the Queen's Award in 2001, 100% organic and biodynamic, Cook's Delight of Berkhamsted, Hertfordshire (run by Rex Tyler) has even cleansed its shelves of any organic products owned by companies linked to GM foods and chemical production, and those whose parent companies are linked to hydrogenated fat production. Cook's also has an online mail order service, which couriers all products (except chilled or frozen) anywhere on mainland UK. Phone 01442 863 584 www.organiccooksdelight.co.uk info@cooksdelight.co.uk.

Fresh & Wild Fresh & Wild healthfood stores/mini organic supermarkets and deli/juice bars are currently the leading chain in London and are now starting to spread across the capital, with the original store at 196 Old Street, EC1 (020 7250 1708), a store in Camden (49 Parkway, NW1, 020 7428 7575), one in Clapham Junction (305–311 Lavender Hill, SW11, 020 7585 1488), Soho (69–75 Brewer St, W1, 020 7434 3179), another in Stoke Newington (32–40 Stoke Newington Church St, N16, 020 7254 2332) and also Notting Hill (210 Westbourne Grove, W11, 020 7229 1063). Fresh & Wild also offers a home delivery service to 72 postcodes around London. Delivery costs £6, or free on orders over £150. Phone 020 7792 9020 (head office) 0800 0322 3456 (home delivery) www.freshandwild.com.

Green Baby* This is a one-stop baby shop at 345 Upper Street, Islington, which stocks washable nappies, babywear, bedding, toys and other organic baby products. The shop is open Monday through Saturday, 10am to 5pm. Also offers mail order. Phone 0870 240 6894 (info), 020 7359 7037 (shop) www.greenbabyco.com info@greenbaby.co.uk.

Infinity Foods Brighton's organic wholefood store and bakery is at 25 North Rd, open six days a week. Also mail order on orders over £300. Phone 01273 603 563. www.infinityfoods.co.uk.

Neal's Yard Remedies Seven shops in London, seventeen shops around the UK and four overseas shops make up the Neal's Yard Remedies chain founded by Romy Fraser and selling the famous little blue bottles of toiletries and herbal remedies (either certified organic or made with as many organic ingredients as possible), books, gifts and more. Phone 020 7627 1949 (customer services) www.nealsyardremedies.com cservice@nealsyardremedies.com.

The Nutri-Centre Based downstairs in The Hale Clinic at 7 Park Crescent, London, the Nutri Centre has a large supplements shop and bookstore used by and recommended by many leading alternative practitioners. Nearest tube stations are Regents Park and Great Portland Street; opening times are Monday to Friday 9am–7pm and Saturday 10am–5pm. Phone 020 7436 5122 www.nutricentre.com customerservices@nutricentre.com.

Olivers Wholefood Store Olivers is a small organic supermarket with knowledgeable health advisers on hand. Large produce and bodycare sections are matched by a floor to ceiling range of organic and biodynamic groceries, chilled and frozen goods, wines and juices. Olivers runs a leading health lecture series, has visiting natural practitioners and its own in-store newsletter. Find it on Station Approach, Kew Village in West London. Phone 0208 948 3990 everyone@oliverswholefoods.co.uk.

Out of this World The UK's leading environmentally friendly and ethical co-operative supermarket chain, with branches in Newcastle-upon-Tyne and Nottingham. Even the counter tops are eco-friendly—made from recycled plastic bottles, yoghurt pots and coffee pots. Annual membership for non-regular shoppers is £12.50 (plus the lifetime membership of £5, for which you receive shares in the company), which includes a national magazine *World News*, many discounts and offers; mail order for some items is available through Natural Collection* catalogue. Find the Newcastle upon Tyne store at Gosforth Shopping Centre, High Street (0191 213 0421) and Nottingham store at Villa Street, Beeston (0115 943 1311). Home delivery is available via LPG-fueled van from the Newcastle-upon-Tyne shop. Phone 0191 213 5377 (info or to join) www.outofthisworld.coop info@ootw.co.uk.

Planet Organic Highly commended in the Organic Food Awards in 1999 and 2000, Planet Organic (run by Renée Elliot) has two large organic supermarkets in London: 42 Westbourne Grove, W2 (020 7727 2227) and 22 Torrington Place, WC1 (020 7436 1929) As well as shelves of thousands of organic and natural products (including the Baby Organix, Babynat, Ecobaby and HIPP ranges), Planet Organic stocks a large range of organic toiletries, vitamin, mineral and superfood supplements, organic wines, meats and fish; it also has a juice bar, and sells fresh produce and books. Many allergy-free foods are stocked, and a homeopath and naturopath are available for consultations five days a week. Home delivery is available for central Londoners for £6.50 on orders less than £100. Open Monday to Saturday 9am–8pm and Sunday 11am–5pm. Phone 020 7221 7171 (head office) 020 7221 1345 (home deliveries) www.planetorganic.com deliveries@planetorganic.com.

Revital Health Stores There are five Revital healthfood stores open six days a week, or you can order a large range of supplements and books etc (over 10,000 lines) over the phone or by email. Orders taken by phone or fax are despatched within 12 hours. In London they are at 35 High Rd, Willesden NW10 and at 3a The Colonnades, 123/151 Buckingham Palace Rd, SW1. Also in Ruislip, Stratford-upon-Avon and Beaconsfield. Phone 01895 629950 (customer services) 0800 252 875 (mail order) www.revital.com sales@revital.com.

LEADING MAIL ORDER SUPPLIERS

21st Century Health P&p costs £3. Phone 020 7935 5440 (info) or 0800 026 0220 (to order), info@21stcenturyhealth.co.uk. or order online at www.aquaball.com.

Beaming Baby Free catalogue available. Order by phone or online. P&p is £3.95, free on orders over £80. Delivery within four working days. Phone 0800 0345 672 www.beamingbaby.com.

Bio-D Direct The major UK manufacturer of environmentally friendly cleaning agents, recommended by the Breakspear Hospital. P&p is free on orders over £25. Order by phone. Phone 01482 229 950 www.bio-dltd.co.uk sales@ecodet.karoo.co.uk.

Born Website has information explaining each different type of nappy system. Also sells slings, nursing bras and breastpads, books and Lansinoh lanolin for breast-feeding. Order by phone or online. P&p starts at £1.47. Phone 0117 924 5080 www.first-born.co.uk info@borndirect.com.

Centre for Alternative Technology This educational eco-centre has a large mail order catalogue with books on every aspect of green living, solar and wind power kits, gardening supplies, building materials as well as Greenfibres organic bedding. They even have a wedding list. Order by phone, post, email or online. Phone 01654 705959 www.cat.org.uk mail.order@cat.org.uk.

Clearspring Direct You can now mail order many excellent organic, macrobiotic, sugar-free and dairy-free groceries from this importer and wholesaler, such as the Clearspring range of Mitoku Japanese Udon, Soba and Ramen wholegrain noodles, Arame and Hijiki Sea Vegetables, and Westbray red and white miso foods. These nutritious wholefoods require little cooking and can be used for older babies and the rest of the family in vegetable meals. (They are also found in leading healthfood stores.) There are also organic fruit purées in four flavours which could be used for babies, and organic unrefined cold-pressed oils. P&p is £5 on orders under £45; free next day delivery for orders over £45. Orders over £100 get a 5% discount. Next day delivery if ordered before 2pm. Order by phone, post or online. Phone 020 8746 1718 (main office) 0871 8716611 (mail order) www.clearspring.co.uk Mail order is handled by Goodness Direct www.goodnessdirect.co.uk.

Clothworks Their founder, mother Linda Row, uniquely designs and makes organic cotton adults' and children's clothes, with organic fibres from Europe and the US. P&p: £1.25 for up to £25-worth of orders, £3.50 for goods ordered from £25 to £60; for orders over £60, p&p is free. Goods are delivered within 10 days. Order by phone or mail. Phone 01225 309218 www.clothworks.co.uk clothworks.info@virgin.net.

Cotton Comfort This allergy-free clothing company charges £3.50 p&p. Order by phone, post or online. Phone 01524 730093 www.eczemaclothing.com.

Cuddlebabes Order by phone, post or email. P&p is 70p for lightweight orders, £1.50 for orders up to £15 and £3 for orders up to £50. Over £50 there is no P&p. Phone 01430 425257 www.cuddlebabes.co.uk.

Eco-babes An Essex one-stop nappy company run by mother Dyane Cakebread of Cambridgeshire. P&p is £2.50 up to £30, £4.50 up to £200 and free for orders over £200. Next day delivery costs £8. Order by phone or online. Phone 01353 664941 www.eco-babes.co.uk info@eco-babes.co.uk.

Garthenor Supply hand-knitted organic pure wool clothing, and also fleece and knitting yarns from many varieties of sheep, including Black Welsh Mountain, Cotswold and Hebridean. Order by phone, online or by post. Phone 01570 493347 www.organicpurewool.co.uk garthenor@organicpurewool.co.uk.

Gossypium Ethical eco-cotton store. Catalogue and online ordering. Delivery within seven working days. Order by phone or online. Phone 01273 897509 (info), 0800 085 6549 (mail order), catalogue@gossypium.co.uk (for free catalogue) or customercare@gossypium.co.uk www.gossypium.co.uk.

Green Baby A one-stop baby shop at 345 Upper Street, Islington, London N1 as well as mail order. Order by phone, post or online. Phone 020 7226 4345 www.greenbabyco.com.

Green People Concerned mother Charlotte Vohtz, originally from Denmark, set up this UK company after finding a curing herbal tonic for her daughter who suffered terribly from eczema. 10% of net profits are donated to charity. P&p costs £2.50 on orders up to £30, £4.50 on orders £30 to £100, free over £100. Order online or by phone or post. Phone 01444 401444 www.greenpeople.co.uk organic@greenpeople.co.uk.

The Green Shop This excellent store with a mail order catalogue is based in Bisley, near Stroud. It specializes in organic paints, renewable energy systems and other household products. Order by phone, post or online. Phone 01452 770629 www.thegreenshop.co.uk.

Greenfibres Run by parents William and Gaby Lana of Devon. P&p is £3.50, with delivery free for orders over £60. An express delivery service is available for an extra £5 subject to stock availability (orders placed by 11am can be received the next working day). Order by phone, post or email. Greenfibres also has a shop in Totnes, Devon, open Monday through Saturday, where you'll find their full range plus some one-offs and Dr. Hauschka products. Phone 01803 868001 (shop) 0845 330 3440 (to order) www.greenfibres.com mail@greenfibres.com.

The Healthy House P&p costs £3.95. Goods usually received within 7 days. Order by phone post or online. Phone 01453 752 216 www.healthy-house.co.uk info@healthy-house.co.uk.

Natural Child Nappies and accessories, Earth Friendly Baby, Green Baby and AromaKids products. Gift packs for mums and babies. Yoga for pregnancy video. Order online or phone/email for catalogue. P&p costs £3 for orders under £60. Delivery within 3 days. Phone 023 9252 6107 www.naturalchild.co.uk info@naturalchild.co.uk.

Natural Collection This is a national mail order catalogue of organic and environmentally friendly goods. P&p is £3.75 on orders up to £100, otherwise free. Delivery within seven working days. Phone 0870 331 3335 (customer services) 0870 331 3333 (to order) www.naturalcollection.com info@naturalcollection.com.

Natural Woman Sells Maggie's organic underwear, organic bed linens, natracare, Earthy Friendly Baby products and a range of other natural products for women. Order online by post, fax or phone. Phone 0117 968 7744 www.natural-woman.com.

Neal's Yard Remedies Phone customer services on 0207 627 1949 to find your nearest store. Mail order is available for £4 on orders up to £100, free over £100. Order by phone, post or online. Phone 0161 831 7875 (to order) www.nealsyardremedies.com cservice@nealsyardremedies.com.

The Nice Nappy Company In addition to nappies and baby clothes, this company stocks a number of products for mothers. It also sells second-hand baby goods and creations by WHAMs (work-at-home mums). Order by phone, email or post. P&p cost £2.50 on orders under £20 and £4 on orders £21-£50. P&p is 10% on orders over £50 up to £100. Delivery within 28 days. Phone 07941 412003 www.nicenappy.co.uk info@nicenappy.co.uk.

Organico Realfoods This distributor of imported French brand Babynat can provide a full technical guide on both its formula milks. For first orders Organico will deliver just one tin for you to try out (for £1.50 p&p) if you don't want to commit to a whole case (6 x 400gm tins). Mail order available: delivery costs £5; free if ordering products worth £50 or more (the minimum order is £20). Order by phone. Phone 0118 951 0518 www.organico.co.uk info@organico.co.uk.

Revital Phone 01895 629950 (customer services) 0800 252 875 (mail order) Order by phone or online. www.revital.com enquire@revital.com.

Savant Distribution Phone for a catalogue of this company's extensive range of organic food and health products. Order by phone or online. Maximum £5 delivery charge. Phone 0845 0606 070 www.savant-health.com.

Schmidt Natural Clothing Based in East Sussex, Schmidt has been selling organic clothing for families for nine years and offers one of the largest ranges. P&p on its nappies is 80p per item up to a maximum of £4. Order by phone or post. Phone 01825 714676 www.naturalclothing.co.uk glenn@naturalclothing.co.uk.

Spirit of Nature Postage is £3.95 on orders up to £60, and free over £60; delivery is within 3–5 days. Order by phone, post or online. Phone 0870 725 9885 (customer services) 01525 381343 (to order) www.spiritofnature.co.uk mail@spiritofnature.co.uk.

Twinkle, Twinkle Wide selection of nappies and accessories, organic baby clothes and toiletries, slings and products for mum as well. P&p costs £1.20 on orders under £15, £2.55 on orders between £15 and £50. Order by phone or online. P&p is free on orders over £50. Orders dispatched in two working days. Phone 0118 934 2120 www.twinkleontheweb.co.uk.

Xynergy Health Products A company offering many supplements, and food and health books. P&p costs £1.95 on orders under £35; over £35 it's free. Order by phone, online or by email. Phone 01730 813642 www.xynergy.co.uk orders@xynergy.co.uk.

ORGANIC FOOD DELIVERY SERVICES

There are now over 400 organic box and home delivery schemes available throughout the UK. Consult *The Organic Directory* to find ones close to you. You may like to try some of the major nationwide organic delivery services listed below. Alternatively, many healthfood stores now offer organic vegetables that you can order, pay for and pick up from the store on a particular day. Larger organic stores stock their own year-round or can put you in touch with a local organic box scheme. Some schemes now deliver wines, chocolates, juicers, cleaning agents, books and even organic clothing and are listed below—some include delivery charges in the price of goods, while others charge extra. However, remember that it is better to buy your food locally from a local farm, box scheme or shop wherever possible, to reduce food miles and support local employment.

Abel & Cole This London based delivery company supplies 100% organic fruit and vegetable boxes nationwide. Boxes are delivered on a weekly basis all over the UK (deliveries are on different days according to where you live). Abel & Cole does an Organic Weaning Box, which include produce suitable for making babyfood, such as bananas, broccoli, carrots, apples, pears and parsnips. Delivery charges are included in box prices. The company has a weekly newsletter and delivers in returnable boxes. Phone 020 7737 3648 www.abel-cole.co.uk organics@abel-cole.co.uk.

Farm-a-round One of the UK's largest suppliers of organic food, delivering to households within the M25, as well as to wholesale customers including hotels, shops and caterers. Fruit and vegetables are the mainstay. Phone 0207 627 8066 Order by phone or online. www.farmaround.net homedelivery@farmaround.co.uk.

The Fresh Food Co. Nationwide organic food delivery service selling fresh fruit and vegetables, fish, meat, dairy products, bread and grocery items. A 'Classic' vege-box contains 6–8kg of seasonal organic produce and costs £26.95. They claim that 70% of the products they sell are grown and produced in the UK. Subscribe for a weekly or fortnightly delivery. Phone 020 8749 8778 www.freshfood.co.uk organic@freshfood.co.uk.

Goodness Foods Organic food delivery service with online ordering only. Baby foods available include Baby Organix, Babynat, HIPP and Nanny's Goat Milk. Phone 0871 871 6611 www.goodnessdirect.co.uk info@goodnessdirect.co.uk.

Graig Farm Organics Voted best organic mail order service by the BBC *Good Food Magazine* in January 2003, this nationwide organic food delivery service based at Graig Farm sells meat, groceries, produce, drinks, dairy, fish, toiletries, pet food, etc. Also offering Truuuly Scrumptious Fresh Baby Food, suitable from four months. Varieties come in 100g resealable plastic containers for £1.10 and include butternut squash, potato, courgette and pea, carrot, sweet potato, apple and pear, and apple. Graig Farm also sells organically grown, biodynamically tanned wool fleeces for £68.92 and an organic woollen baby blanket for £15.24. Many healthfood shops sell Graig Farm produce. Order direct online, by phone, fax, post or email. Delivery is free on orders over £45. Meat is delivered in returnable polystyrene boxes. Phone 01597 851655 www.graigfarm.co.uk.

Healthy Hampers An all-year-round vegetarian and organic foods gift service with nationwide delivery. P&p included, prices start from £49.95 including a wicker basket with message of choice. Phone 01252 878 698.

My Organic Online organic shopping source. Sells baby food including Babynat, HIPP and Eco-baby muesli. Phone 020 7436 8066 www.myorganic.co.uk info@myorganic.co.uk.

The Organic Delivery Company An evening delivery of fresh fruit and vegetables from small farmers and growers to London postcoded homes plus Richmond and Kingston. Free delivery on orders of £25 or over. Boxes can be reused—the driver will pick up with your next order. Order online or by phone. Phone 020 739 8181 www.organicdelivery.co.uk.

The Organic Shop Nationwide delivery company selling fresh fruit and vegetables, fish, meat, wine, dairy and bakery products, household goods and grocery items including Baby Organix. This company donates 10% of its profits to charity. Delivery free on boxes over £25 or wine orders over £75. Delivery in about two days. Phone 0845 674 4000 www.theorganicshop.co.uk info@theorganicshop.co.uk.

Riverford Organics Southern England home delivery of a wide range of organic vegetables, fruit and locally produced organic foods such as eggs. Riverford is the largest independent producer of organic vegetables in the UK and has two farm shops in Devon. Mail order customers are asked to return containers— drivers will collect them on next delivery. Phone 01803 762 720 (mail order) www.riverford.co.uk.

Simply Organic Nationwide delivery company selling fresh fruit and vegetables, fish, meat, dairy products and grocery items including Baby Organix and HIPP babyfoods. They also sell Weleda baby toiletries and Tushies wipes. Delivery costs £5. You receive a £5 gift voucher for setting up a standard order. Phone 0845 1000 444 (to order) 01604 791911 (other enquiries) www.simplyorganic.net orders@simplyorganic.net.

Swaddles Nationwide organic delivery service selling fresh fruit and vegetables, meat, fish, dairy products and grocery items including organic baby and children's food. Minimum order of £25. Delivery charge is £10 on orders under £50, £5 on orders between £50 and £100, free on orders over £100. Phone 0845 4561768 www.swaddles.co.uk information@swaddles.co.uk.

The Village Bakery The largest range of organic baked goods available by mail order is supplied by The Village Bakery. Choose from scrumptious, wholesome organic breads, biscuits, cakes, jams, flours, slices and Christmas goodies. Many of the foods are suitable for those on special diets and are free of sugar, gluten, wheat, yeast and dairy. P&p is £5.95. Order online or by phone. Phone 01768 881515 www.village-bakery.com info@village-bakery.com.

WEB RESOURCES:
Useful websites and search engines

Baby World www.babyworld.co.uk An online search engine including which organic baby foods are available in other countries. They sell some organic books and have an online shop selling some organic foods.

Food First www.foodfirst.co.uk Web search engine for both conventional and organic UK producers, suppliers and retailers. Also offers email newsletter.

Food Net www.foodnetuk.com Lists British producers of meat, baked goods, dairy, fish, fresh produce, drinks, etc. Also includes nationwide list of farm shops.

Green Choices www.greenchoices.org Online information source for all things green, including links to baby relevant sites.

Green Guide www.greenguide.co.uk Searchable website for 'green' travel destinations, food and drink.

International Organic Cotton Directory www.organiccottondirectory.net Lists organic cotton producers, manufacturers and retailers around the world.

Links Organic www.linksorganic.com International search site for all things organic.

The Organic Directory www.theorganicdirectory.co.uk The most comprehensive listing of organic retailers, wholesalers and other suppliers, hosted on the Soil Association's website.

Organic Food www.organicfood.co.uk Online UK organic magazine with articles, book reviews, chat zone and a box scheme directory.

Organic Supermarket www.organic-supermarket.co.uk Searchable directory of organic and natural food and goods suppliers in the UK, including baby food.

Kid's Organic Club www.kids.organics.org Competitions, games (including Fun Farm activity site), a cast of characters and explanations of how animals are reared on organic farms can be found on this website by Pure Organics, who also manufacture a range of children's meat and vegetarian organic frozen foods.

PUBLICATIONS

Many of the following books and magazines have been used as references, information and motivation for the author's parenthood, work as a journalist, and for *The Organic Baby Book*. They are ideal for delving into interesting subjects in more depth, keeping up to date with issues, news and developments. There are also a few of the many appropriate children's books now available to interest your little ones in a more environmentally friendly way of life.

You can get many of these books from independent retailers such as the Nutri-Centre Bookshop (see the Mail Order Services list on page 177). If out of print, books can often be easily obtained over the internet through Amazon and other second-hand internet booksellers.

Magazines & Journals

Baby G.R.O.E. This stands for Green, Recycled, Organic, Ethical: a new free brochure with advice for parents wanting to raise their children in an ethical, green way. Includes information on and advertisements from many leading organic, green and ethical companies. info@babygroe.co.uk.

BBC Good Food Magazine Features many new organic products, food news, views, features and information. Available in newsagents for £1.85 or by subscription for £23 for 12 issues with many discounts and offers. Phone 01795 414712 www.bbc.co.uk/food/.

Ergo A funky magazine covering ethical consumer issues. Phone 020 7405 5633 www.ergo-living.com.

Ethical Consumer Magazine Published six times a year by the Ethical Consumer Research Association. A detailed read, this magazine researches and compares consumer goods, company claims and includes news of product boycotts, etc. Available on subscription. Phone 0161 226 2929 www.ethicalconsumer.org.

Food Magazine Quarterly news magazine of the Food Commission, a non-profit organisation which campaigns for the right to safe, wholesome food—the UK's leading independent consumer watchdog on food standards. Edited by Sue Dibb and Tim Lobstein, it contains exposés on food and health issues, news items and reviews of new consumer foods, food company activities and food and health books. Its stories are often picked up by the national press. Available on subscription for £20. Phone 0207 837 2250 www.foodcomm.org.uk info@foodcomm.org.uk.

Green Events News, events, courses, listings and useful ads in these monthly newsletters. Versions available in London, Devon and other regions. www.greenevents.fsnet.co.uk (London and Southeast) www.greenevents.co.uk (Devon and Southwest).

Here's Health Glossy consumer magazine which contains the most organic, holistic and environmentally friendly news, information, products and features of all the glossies. Subscription is £21.60 for 12 issues. Phone 01858 438861 www.emapmagazines.co.uk subs@subscription.co.uk.

The inside story A small, hard-hitting bimonthly newsletter for trade, health professionals and media, collecting food and health issue bytes with the emphasis on allergies by journalist Michelle Berriedale Johnson. Features news, special diets, recipes, articles, reviews and product evaluations. Subscription costs £28.95 a year. Phone 0207 722 2866.

Living Earth The quarterly membership magazine of the Soil Association, free to members (joining fee is £24 per year). Covers interesting organic news and issues with feature stories and new products. Available in some healthfood stores or from the Soil Association. Phone 0117 929 0661 www.soilassociation.org info@soilassociation.org.

New Consumer Similar to *Ethical Consumer*, but has more focus on the fashionable. It costs £22.50 for the first six issues. Phone 0800 389 4728 newconsumer@bigissuescotland.com.

Optimum Nutrition—Journal of the ION A biannual journal from the Institute for Optimum Nutrition, with news, research updates, details of ION workshops, food facts, recipes, book reviews, new products and features. It costs £3.50 or comes free with annual Club ION membership of £24. Phone 020 8877 9993 www.ion.ac.uk info@ion.ac.uk.

Organics Bimonthly magazine covering all things organic, put out by the wViP group. Call or email wViP for information and subscriptions. Phone 020 7331 1000 organics@wvip.co.uk.

The Organic Way—the HDRA Magazine This quarterly magazine comes free with membership of the Henry Doubleday Research Association, offering news, product and relevant book reviews, handy gardening tips, your chance to ask the experts about gardening problems, and much more. To join and subscribe to the magazine at the same time costs £25 for a family, £21 for an individual, and £14 for concessions. Phone 02476 303 517 www.hdra.org.uk enquiry@hdra.org.uk.

Permaculture Magazine News, views, features and more on permaculture (sustainable food and lifestyles, environmental issues etc) are in this quarterly magazine from Permanent Publications. It costs £10 a year to subscribe, and back issues are available. Permanent Publications also carries over 500 books, CDs, videos, etc. in their Earth Repair Catalogue. Phone 0845 45841450 www.permaculture.co.uk info@permaculture.co.uk.

Pesticide News Quarterly journal of The Pesticide Action Network (PAN-UK) with news, research snippets, information on new regulations, book reviews and resources. Subscription is from £20 for individuals and there are also stand alone report publications such as *Organic Cotton—from field to final product* (£16.50). Phone 020 7274 8895 www.pan-uk.org admin@pan-uk.org.

Positive News & Living Lightly Magazine and popular newspaper covering positive developments in the eco-spiritual world, environmental events and information. Subscriptions are £12.50 a year for both publications. Phone 01588 640022 www.positivenews.org.uk office@positivenews.org.uk.

Proof! A quarterly newsletter which road tests remedies, seeks out scientific evidence for therapies, and explores best treatments for specific health problems. From the editors of *What Doctors Don't Tell You*, this newsletter costs £34.95 for four issues. Phone 0870 444 9886 www.wddty.co.uk cs@wddty.co.uk.

Pure Published by Green Guide, this monthly organic magazine costs £2.95 per issue from health food stores and some book stores. When ordering 12 issues online you receive a 20% discount off the retail price. Phone 020 8883 0933 www.greenguide.co.uk.

What Doctors Don't Tell You A factual monthly newsletter giving you and your family latest information about health issues, illnesses and treatments based on scientific evidence around the world—facts that the editors claim the medical establishment and drugs companies would rather you didn't know. Subjects are covered in depth, and there are news items, medical case studies, alternative medicine features, etc. Twelve issues costs £42.79 with discounts on conferences and books. Phone 0870 444 9886 www.wddty.co.uk cs@wddty.co.uk.

Publications: Parenting, food & health

Baby Wisdom Customs and traditions of baby-raising from around the world. Covers the first year, including how parents from different cultures deal with sleep problems, breastfeeding, naming, etc. By Deborah Jackson and available from Born* for £14.99.

Everything There is to Know about Vaccination: The Essential Guide for Parents Compiled by Joanna Karpasea-Jones of the Vaccination Information Service, this details parents' rights, study results on vaccination use and the ingredients used in vaccines (aluminium, mercury, sodium hydroxide etc) and the tissues they are cultured in (including monkey kidney and egg). Available for £4.95 from the Vaccination Awareness Network (VAN-UK). Phone 0870 4440894 www.vaccine-info.com enquiries@vaccine-info.com.

The Food our Children Eat by leading investigative journalist Joanna Blythman, published by 4th Estate. This book shows how children are eating the worst food rather than the best that they need for healthy growth and development, and features useful strategies and recipes to get them back on track. £7.99 from bookstores and mail order catalogues.

Foresight: Planning for a Healthy Baby by Belinda Barnes and Suzanne Gail Bradley. Essential reading for all future organic parents—follows the Foresight programme and successes. Available from Foresight for £9.99 including p&p. Phone 01483 548071 www.foresight-preconception.org.uk/home-page.html.

Grow Organic Eat Organic: A practical activity book for beginners by Lone Martin. An activity book showing how food gets from soil to table, price £4.99.

Having Faith: An Ecologist's Journey into Motherhood This book by Sandra Steingraber delves into the pollutants to which babies are exposed from conception onward, from a very personal perspective. It costs £11.99.

Healthy Parents: Better Babies: A Couple's Guide to Natural Preconception Health Care by Francesca Naish and Janette Roberts, published by Gill & Macmillan £12.99.

The New Natural Pregnancy Available from the Active Birth Centre for £8.99, the book covers the appropriate use of holistic therapies during pregnancy including homeopathy, acupuncture, aromatherapy, hypnotherapy, yoga, nutrition and massage. Published by Gaia Books.

Organic Baby & Toddler Cookbook by Tanyia Maxted-Frost and Daphne Lambert and published by Green Books in 2000, this companion to *The Organic Baby Book* offers a nutritional guide for weaning your baby to toddlerhood. Promotes an organic, seasonal, mainly raw wholefood diet including cakes and snacks. Available from independent healthfood stores and Green Books for £6.95 (plus £1.50 p&p).

The Organic Directory The most comprehensive listing of organic retailers, wholesalers, importers, etc etc. 2002/3 edition 384 pages, price £4.95. Phone 01803 863260 www.greenbooks.co.uk sales@greenbooks.co.uk.

Preparation for Pregnancy: an essential guide A guide for practitioners by Suzanne Gail Bradley and Nicholas Bennett. International research into pre-conceptual health and pregnancy outcome—the Foresight experience. Available from Foresight for £8.99 including p&p. Phone 01483 548071 www.foresight-preconception.org.uk/home-page.html.

Super Baby Boost your baby's potential from conception to year one. By Dr Sarah Brewer, published by Thorsons in 1998, price £7.99. Very informative, featuring technical and practical information, including many useful tips on stimulating your baby in the womb, and mentally and physically enriching him in his first year of life. Out of print.

Wise Woman Herbal for the Childbearing Year This book by Susan Weed covers many commonly available herbs and their appropriate uses for pregnancy, childbirth and your baby's first months. Costs £6 from bookstores and mail order booksellers.

For further details about companies marked with an asterisk (*) including how to order, postage & packing charges etc, see the Leading Mail Order Suppliers section on pages 164–167.

Publications: The Environment

Clean House, Clean Planet: A manual to free your home of 14 common hazardous household products This American book by Karen Noonan Logan explains about the hazards of commercial cleaners, and teaches you how to make your own simple and safe ones. Her recipes are very easy to follow and effective. Published by Pocket Books in the US, available in the UK for £8.75 approx.

Eco-Friendly Houseplants by B. C. Wolverton, this book describes and rates common household plants and their ability to remove certain toxins from the air including formaldehyde and other VOCs. Available from gardening shops and bookstores for £9.99.

The Feminisation of Nature by Deborah Cadbury, an Emmy Award-winning producer of science programmes for television. She examines the evidence that alarming changes are taking place in human reproduction and health: sperm counts are falling, testicular, prostate and breast cancer are on the increase, and different species are showing signs of 'feminisation', or even changing sex. The prime suspect is the increased exposure to chemicals which mimic oestrogen and other hormones. Out of print.

Genetic Engineering, Food and Our Environment An excellent, inexpensive guide to the issues by activist Luke Anderson. Includes a wealth of information, backed up by extensive references. Available in bookshops and by mail order for £3.95 plus £1.50 p&p from the publishers Green Books.

The Green Guide An annual comprehensive environmental consumer directory for various parts of the UK, including London, Wales, Scotland and the Southwest. Also a Christmas Green Guide. Sold in healthfood stores or order from Green Guide directly. Also offers an mail order catalogue similar to Natural Collection. Phone 020 8883 0933 www.greenguide.co.uk sales@greenguide.co.uk.

How to Avoid GM Foods A practical shopper's guide by leading investigative journalist Joanna Blythman, published by 4th Estate in 1999. Names and shames the companies and products containing GM ingredients. The UK edition is out of print, but the US edition is available through Amazon, price £7.00 approx.

The Humanure Handbook For the really committed reader! The Chinese have recycled human manure for thousands of years, and this book by Joseph Jenkins shows how you can do it. Available from Green Books, revised 1999 edition, £13.95.

Increase Your Sex Drive Eating organic food is just one way to spice up your sex life. A practical book by Dr Sarah Brewer on the amazing power of supplements, herbs, instant aphrodisiacs and foods which can moderate your sex hormones. Published by Thorsons for £7.99.

Natural Product News The monthly leading organic and natural products magazine in the UK, aimed at the trade sector with industry news, developments, product and issue features. Available on subscription from Full Moon Communications, who run the annual Natural Products trade exhibition. Phone 01903 817 303.

Organic Gardening books—main sources There are many excellent books to choose from on organic gardening. Get hold of the catalogues from the Henry Doubleday Research Association and Green Books, and The Earth Repair Catalogue from Permanent Publications, and decide which ones are best for you.

Our Stolen Future The true, shocking story by scientists Theo Colborn, Dianne Dumanoski and John Peterson Myers, about the toxic chemical legacy being passed from generation to generation. Published by Abacus, price £8.99, and available through some mail order catalogues. Considered by many as a worthy successor to Rachel Carson's *Silent Spring*.

The Perils of Progress The health and environmental hazards of modern technology, and what you can do about them are investigated by this in-depth book, which looks at the dangers in the food we eat, electrical appliances, water quality and more. By John Ashton and Ron Laura, published by Zed Books in 1999, price £14.95.

Recycled Products Guide This useful guide is now available online from Waste Watch. It's a comprehensive listing of recycled materials, manufacturers and suppliers. Includes clothing, garden furniture, paper, plant pots, soil, stationery etc. www.wastewatch.org.uk (general info) www.recycledproducts.org.uk (online guide).

Silent Spring This book is widely claimed to have begun the modern environmental movement and to be one of the most influential of the century. Written by scientist Rachel Carson of the US, it tells the shocking truth behind 'progress' which poisons the environment, wildlife—and us. It is still, unfortunately, extremely relevant for today. Originally published by Penguin Books in 1962, still available as a Penguin paperback, price £8.99.

Publications: Food, nutrition & cookery

Family Organic Cookbook Another cookbook from Carol Charlton, which includes strategies for how to get children to eat vegetables and healthy snack ideas. Around £18.99 from bookstores and mail order booksellers.

The New Shopper's Guide to Organic Food by journalist Lynda Brown, published by Fourth Estate, 1999, £9.99. The previous edition was the first mainstream organic book in the UK encouraging consumers to go organic, with many interesting facts and figures and useful organic shopping information for consumers. Available in leading bookstores and from the Food Commission. Phone 0207 837 2250 www.foodcomm.org.uk info@foodcomm.org.uk.

The Optimum Nutrition Bible All you need to know for optimum nutrition by Patrick Holford—the bible for health fanatics which has been reprinted many times. Available from leading bookshops and mail order catalogues. Published by Piatkus, 1997, £12.99.

The Organic Café Cookbook A stylish organic cookbook with recipes, features on the people who grow the food, and a few facts thrown in for good measure. Carol Charlton of The Organic Café in Queens Park, London, will convert you to organic if you're still on the fence, or lead you on an exciting culinary journey. Published by David & Charles, 1999, £18.99.

The Organic Cookbook: Naturally Good Food by Renée Elliot of Planet Organic and Eric Truille, the US edition of this book is still available, price £10 approx.

Organic: Recipes for a New Way of Eating A comprehensive overview of organic food, with many recipes, by Sophie Grigson and William Black, price £25.

The Organic Directory Find suppliers of organic food and many useful organisations and organic contacts throughout the UK in this directory by Clive Litchfield. Published by Green Books, 2002/3 edition price £4.95.

Organic Living by Michael van Straten, £19.99.

Planet Organic: Organic Living This guide to living organically by leading journalist Lynda Brown costs £14.99.

Raw Energy series *The Raw Energy Bible* and *The New Raw Energy* are just two of Leslie Kenton's excellent dynamic food and health books, published by Ebury Press and available from many bookstores and mail order catalogues. Mostly priced at £6.99.

Super Juice: Juicing for health and healing with detox juice diet, shakes and smoothies, power and vitality juices—the essential guide. By naturopath Michael van Straten, published by Mitchell Beazley in 1999, price £7.99. Out of print.

'Super' series by Michael van Straten, published by Mitchell Beazley. The series includes *Super Boosters: herb, plant and spice extracts to boost health* (£10.99), *Super Soups, Super Salads, Super Radiance Detox* and *Super Energy Detox*.

Children's books on green issues

The Gigantic Turnip A classic Russian tale with lovely illustrations which won the 1999 Books for children Mother Goose Award. This hardback from Barefoot Books—featuring all the farmyard animals—costs £9.99 and is suitable for ages 2–6.

Herb, The Vegetarian Dragon A lively picture book which will interest children from around one year onwards, due to the vibrant, contrasting and interesting illustrations and adventurous story. Herb grows his own vegetables and brings peace to the mediaeval world with a little help from a brave girl—uniting dragons, knights, royalty and rabble. The book costs £4.99. *Cooking with Herb*, a cookbook for children and companion to the storybook, costs £6.99. Both from Barefoot Books.

The Little Red Hen and The Ear of Wheat A classic tale is retold with lovely illustrations of the little red hen who makes a loaf of bread without help from her lazy friends—teaching children how plants grow and where bread comes from. It is a paperback suitable for two to six year olds and costs £4.99 from Barefoot Books.

Muck & Magic By Jo Readman, published by Search Press for the Henry Doubleday Research Association, £4.95. A colourful, fun, practical book for children on how to garden organically, including how to recycle, collect your own seeds, make a wormery, look after your soil and grow your own fruit for drinks. Available from the HDRA.

Planet of the Bugz A cosmic environmental tale which children from three will love, about a Love Bug spinning kite which comes to life and takes a young girl to the amazing Planet of the Bugz. The bugz were like humans who were overworked and polluted their environment to the point where the planet turned grey and they forgot how to fly. Luckily they were able to reclaim their magic and save their planet—which is their message for Earth. It costs £4.95 from mail order stockists.

Rachel's Roses Another charming illustrated book from Barefoot Books, suitable for two to five year olds. The story of a little girl who discovers the joy of growing her own roses. It costs £4.99.

Mail Order Sources for Publications

Barefoot Books 020 8343 6115 www.barefoot-books.com
Centre for Alternative Technology 01654 705959 www.cat.org.uk
Earth Repair Catalogue 0845 45841450 www.permaculture.co.uk
Gaia Books 01453 752 985 www.gaiabooks.co.uk
Green Books 01803 863 260 www. greenbooks.co.uk
Henry Doubleday Research Association 02476 303 517 www.hdra.org.uk enquiry@hdra.org.uk
Institute of Optimum Nutrition 0208 877 9993 www.ion.ac.uk
Nutri-Centre Bookshop 020 7436 5122 www.nutricentre.com
Revital 0800 252 875 www.revital.com
Xynergy Health Products 0845 658 5858 www.xynergy.co.uk

Many relevant publications are also found in leading stores such as Fresh & Wild, Planet Organic, Neal's Yard Remedies, Olivers and Revital, and are also available through some of the Mail Order Suppliers listed on pages 164–167.

ORGANISATIONS
Parenting & health

The Active Birth Centre An education centre for expectant and new parents, health workers, midwives and baby massage instructors. Offers preparation for childbirth classes, post natal events, water birth resources and a natural therapy clinic, online mail order service as well as a register of certified active birth teachers. Some organic mother and baby toiletries. Phone 020 7281 6760 www.activebirthcentre.com mail@activebirthcentre.com.

Baby Milk Action This organisation promotes breast-feeding around the world and campaigns against unsafe milk formulas. They have a few products including books, posters and a breast-feeding calendar. Phone 01223 464 420 www.babymilkaction.org info@babymilkaction.org.

Foresight—The Association for the Promotion of Pre-conceptual Care A must for anyone planning a baby. Foresight is a registered charity offering membership, a newsletter, video, publications, courses, pre-conceptual health programme and hair mineral analysis service. Send an A5 SAE with a 33p stamp to Foresight, 28 The Paddock, Godalming, Surrey GU7 1XD, to receive a catalogue and general information. Foresight has a video ('Preparing for the Healthier Baby') (£15 incl p&p). You can also request a list of Foresight practitioners. Phone 01483 427839 (main office) 01483 548071 (mail order) www.foresight-preconception.org.uk/home-page.html.

The General Council and Register of Naturopaths Call to get a list of naturopaths in your area. The Naturopathic Helpline is staffed from 9am–1pm and from 2–5.30pm. Phone 0870 745 6984 www.naturopathy.org.uk admin@naturopathy.org.uk.

Gentle Birth Preparation Programme—The Jeyarani Way Developed by London obstetrician Dr Gowri Motha. Incorporates alternative health practices by way of classes or self-help. Phone 020 8530 1146 (books & videotapes) 020 8530 114 (appointments & classes) www.jeyarani.com.

The Informed Parent Vaccination information and support. Membership with quarterly newsletter, reading list and introductions to local groups. Send an SAE to PO Box 870, Harrow, Middlesex HA3 7UW. Phone 0208 861 1022.

La Leche League This international organisation provides information on and support for breastfeeding. Phone 020 7242 1278 www.laleche.org.uk.

National Association of Nappy Services Contact NANS to find contact details of a nappy washing service serving your area. Phone 0121 693 4949 www.changenappy.co.uk info@changenappy.co.uk.

National Childbirth Trust The long-established support organisation and network for parents promoting breastfeeding and providing helpful breastfeeding equipment. NCT branches regularly hold secondhand clothes and toy sales. Phone 0870 770 3236, 0870 444 8708 (breastfeeding line) www.nctpregnancyandbabycare.com.

The Natural Nurturing Network Send an SAE to PO Box 48, Matlock, Derbys DE4 3ZR. The network promotes natural parenting, including long-term breastfeeding and having your baby sleep in bed with you. There's a quarterly newsletter, regional group meetings, events, etc. Membership is £10.

The Natural Parent Circle A new contact network to support parents adopting natural parenting. An alternative National Childbirth Trust. The Circle has teamed up with organic suppliers to offer discounts on organic box deliveries (Organics Direct) and many products such as nappies. There are information packs on long-term breastfeeding, vaccination and other subjects. An annual subscription costs £42.60 for a family, or less if you already subscribe to Natural Parent magazine. Phone 0207 354 4592.

The Parent Network This organisation runs regular courses in England, Scotland and Wales designed to enhance communication between parents and children. Topics covered include feelings, raising self-esteem, setting limits and conflict management between children. Phone 0207 735 1214.

The Primal Health Research Centre This London-based charity is run by Dr Michel Odent, the French natural childbirth pioneer and a leading proponent of pre-conception care and the Accordion Method of losing contaminated fat before conception. The Centre has an online databank (now available to buy on disk) of 400 entries exploring the correlation between the primal period (birth to one year old) and health later in life. The Centre is based at 59 Roderick Road, London NW3 2NP. Call for information and course programme. Phone 0207 485 0095 Modent@aol.com.

Real Nappy Association For a parenting pack and information on Real Nappy Week and real reusable cotton and wool nappies (suppliers and nappy washing services) send a large SAE with two first class stamps to PO Box 3704, London SE26 4RX. Membership is £7 for individuals, which gives you a quarterly newsletter with useful tips and suggestions for real nappy action, plus 10% discount on most nappies. Phone 0208 299 4519 www.realnappy.com contact@realnappy.com.

The Society of Homeopaths Call to get the Register of Homeopaths, search online or send an SAE to the General Secretary, The Society of Homeopaths, 2 Artizan Rd, Northampton NN1 4HU. Phone 01604 621 400 www.homeopathy-soh.org info@homeopathy-soh.org.

Organisations: food & environment

Association of Unpasteurized Milk Producers and Consumers This group campaigns for the right for consumers to have 'real' cow's milk—a living food, untreated, with all its vitamins, minerals, enzymes etc, intact. Some farmers believe that whey proteins, the most nutritious, are denatured by heat treatment, causing a loss in value and possibly triggering allergic reactions. There are still around 150 unpasteurized milk producers left in the UK producing 'green top' milk. For information contact the Hardwick Estate Office, Whitchurch-on-Thames, nr Reading, RG8 7RB. Phone 0118 984 2955.

Biodynamic Agricultural Association of Great Britain UK approved organic certification body using the Demeter biodynamic symbol. Produces its own publication (*The Star & Furrow*) twice a year, and a quarterly newsletter. Membership is £25. The office is open from 9am–1pm. Phone 01453 759 501 www.biodynamic.org.uk bdaa@biodynamic.freeserve.co.uk.

Centre for Alternative Technology (CAT) Europe's leading eco-centre, with an environmentally friendly visitor complex, residential courses throughout the year, publications, a mail order catalogue and a membership organisation (the Alternative Technology Association). Based at Machynlleth, Powys. Phone 01654 705950 (general info) 01654 705959 (mail order) www.cat.org.uk.

The Food Commission A non-profit watchdog organisation which campaigns for the right to safe, wholesome food for all consumers. Current campaigns include an irradiation awareness-raising campaign and The Parents Jury, an independent group of about 800 concerned parents, which judge the worst and best foods targeted at children. *The Food Magazine*, a quarterly magazine, contains exposés on food and health issues, news items and Food Commission surveys and reviews of new consumer foods, food company activities and food and health books. Support the Commission by subscribing to *The Food Magazine* for £20. Phone 0207 837 2250 www.foodcomm.org.uk info@foodcomm.org.uk.

The Fresh Network This enterprising national organisation brings information and support to anyone with an interest in the benefits of eating a mainly or totally raw food diet, which has been shown to improve health dramatically and reduce or eliminate allergies and many other ailments. For a £30 membership you get 12 newsletters, a quarterly magazine (*Get Fresh!*) and a 5% discount on mail order goods including books, juicers, supplements and other products. Phone 0870 800 7070 www.fresh-network.com info@fresh-network.com.

Friends of the Earth One of the two largest international campaigning environmental networks. Their Real Food Campaign advocates for local and organic food and campaigns a against genetic engineering. Find out how you can help by becoming a member and joining a local group. Phone 0207 490 1555 www.foe.co.uk.

Genetic Engineering Network / Genetix Update Subscribe to this leading campaigning group to find out the latest GM info. A £5 donation is requested for a newsletter by post, or by email—the email service brings news every few days, and the bi-monthly printed newsletter is a summary of stories. GEN also organizes and takes part in actions—ask for details. Phone 020 8374 9516 www.genetixaction.org.uk.

Greenpeace One of the largest international environmental campaign organisations. Greenpeace is calling for a ban on GM foods and their production and trials, and their replacement with organic food and farming. Join the campaign and find out what you can do in your local areas to help. Phone 0207 865 8100 www.greenpeace.org.uk info@greenpeace.org.uk.

Henry Doubleday Research Association Based at Ryton-on-Dunsmore near Coventry, the HDRA is Europe's largest organic organisation promoting organic gardening and food consumption. It has model organic gardens at Ryton Organic Gardens with courses and events throughout the year, an organic shop with organic food, books and seeds, and an organic restaurant. There's also Yalding Organic Gardens in Kent, and Audley End Kitchen Gardens in Essex to visit. Membership gives you *The Organic Way*, their quarterly magazine, free advice, 10% discount off goods from mail order catalogues, use of their reference library, and free admission to organic gardens throughout the UK. Phone 02476 303 517 www.hdra.org.uk enquiry@hdra.org.uk.

The Institute for Optimum Nutrition Nutrition diploma courses, short workshops and seminars are run by this charity, founded in 1984 by Patrick Holford, which can also provide a list of its qualified nutritionists and stocks a full range of health books. There's a mail order book service and a resource library with over 600 nutrition and health books, journals and audiovisual material, which is open during working hours five days a week. Join Club ION for £24 to have unlimited access to the resources, receive discounts, and get their biannual journal. The ION also has a clinic in Putney. Phone 020 8877 9993 www.ion.ac.uk info@ion.ac.uk.

National Association of Farmers Markets Directory of farmers markets across the UK. Phone 01225 787914 www.farmersmarkets.net nafm@farmersmarkets.net.

Organic Farmers & Growers UK approved organic certification body. Phone 01743 440512 www.organicfarmers.uk.com.

Organic Food Federation UK approved organic certification body. Phone 01760 720444 Website:www.orgfoodfed.com.

Pesticide Action Network (UK) An independent charity concerned with the health and environmental effects of pesticide use which publishes *Pesticide News* (see publications) and many other useful publications. A large reference base of relevant studies and information is kept and can be accessed. The Network is currently working with other environmental groups to get the organochlorine lindane banned. Phone 020 7274 8895 www.pan-uk.org admin@pan-uk.org.

Rudolf Steiner House & Theatre Workshops and theatrical productions for children and adults, and workshops on biodynamic gardening and personal growth are held at Rudolf Steiner House at 35 Park Rd, London NW1 6WT. There's also a library, a bookshop, a quarterly magazine (*New View*), information on Steiner schools, a lecture series on Tuesday evenings, and biodynamic Weleda beauty products are on sale. Phone 0207 723 4400 www.anth.org.uk/RSH/ rsh@cix.compulink.co.uk.

The Soil Association This is the leading UK charity promoting organic food and farming. The Soil Association, based in Bristol, offers membership (from £24 per year), a quarterly magazine and other publications, an annual programme of events including lectures and conferences, and certification and technical advice for organic processors, manufacturers and producers. It campaigns to have organic farming recognized and supported by the government, and against genetic engineering and the over-use of antibiotics, and publishes an annual industry report on the state of the organic food market (sponsored by Baby Organix). The Soil Association has established an Organic Farms Network to enable the public to see why organic farming is so beneficial for themselves—see Organic Outings for details. They also have a Baby & Toddler Club, where upon joining (£2 per month) you receive *The Organic Baby & Toddler Cookbook*, Baby Organix magazine, a recipe book, and *Living Earth* magazine. Phone 0117 929 0661 www.soilassociation.org info@soilassociation.org.

Triodos Bank One of the country's leading ethical banks, which runs an Organic Saver Account which helps put money back into organic farming by donation from the bank to the Soil Association. Triodos also lends to a lot of organic businesses. Phone 0117 973 9339 www.triodos.co.uk mail@triodos.co.uk.

Women's Environmental Network This campaigning organisation can be contacted at PO Box 30626, London E1 1TZ. WEN promotes Real Nappy Week, offers an online index of Real Nappy Partnerships including local authorities, the NHS and companies, and campaigns on issues which link women, health and the environment. Membership costs £15 for individuals or £10 concessions. Phone 020 7481 9004 www.wen.org.uk info@wen.org.uk.

The Women's Nutrition Advisory Service If you join the WNAS Club, you receive discounts on multivitamin and mineral supplements and more. The service was founded by Maryon Stewart (author of *The Phyto Factor* and *Healthy Parents, Healthy Baby*) fifteen years ago to provide professional advice, support information and publications for women of all ages. There are now four WNAS clinics in London and Sussex, postal and telephone consultations (you complete a health and diet diary first), self-help books such as *Nutritional Medicine* and *The Vitality Diet*, and leaflets. Club members (it costs £20 to join) get discounts on services and products, a quarterly newsletter, free samples and more. There are also telephone line support services with pre-recorded messages on subjects including how to get fit for pregnancy and breastfeeding, overcoming fatigue, beating migraines naturally, and skin, nail and hair signs for deficiency. Also an online shop. Send a cheque for £20 for membership to PO

Box 268, Lewes, East Sussex BN7 2QN or contact WNAS. Phone 01273 487 366 www.wnas.org.uk wnas@wnas.org.uk.

The Wildlife Trusts These groups run events locally throughout the country for children and adults. Contact the Trusts' national office to get details of your local group and information on local nature reserves open to the public. Phone 0870 036 7711 www.wildlifetrusts.org.uk info@wildlife-trusts.cix.co.uk.

Vaccination Awareness Network UK Subscribe to this non-profit volunteer-run organisation founded to provide unbiased information on vaccinations. Join from £10 to get four SHOTS newsletters a year, access to a library of books, audio tapes and videos, and access to a support network. Contact the Vaccination Information Service on 0870 444 0894 www.vaccine-info.com enquiries@vaccine-info.com.

LEADING PRO-ORGANIC PRACTITIONERS & PROGRAMMES

Kitty Campion, the Campion Clinic Herbalist, healer, author and naturopath Kitty Campion runs a seven day or seven week complete guided detox programme using complementary therapies, which can be used by couples during pre-conception preparation. She runs the Campion Clinic at 45 Alchester Road, Chesterton, Oxfordshire, OX26 1UN. Phone 0207 722 9270.

Jan de Vries The renowned naturopath and his team of qualified practitioners in homeopathy, naturopathy, iridology etc practice from the Jan de Vries Healthcare Centre in West London. Phone 020 7262 6069 or 020 7224 8701.

Dr Marilyn Glenville Nutritional therapist Dr Marilyn Glenville is an author and broadcaster, former chair of the Foresight Association for the Promotion of Pre-conceptual Care. For over twenty-five years she has studied and practised Nutritional Therapy in the UK and US, specializing in the natural approach to female hormone problems. The author of best-selling *Natural Alternatives to HRT* and *Natural Alternatives to Dieting*, Dr Glenville practises from Women's Healthcare, a private gynaecological clinic, and also the Hale Clinic in London. Phone 08705 329244.

Hospital of St John & St.Elizabeth The private Birth Unit in St Johns Wood specializes in holistic births, preparation and post-natal classes, homeopathy, water births, antenatal and post-natal yoga, nutritional advice, Reiki, baby massage, reflexology and lots more. Phone 020 7806 4000 (hospital info) 020 7806 4090 (Birth Unit) www.hje.org.uk.

The Hale Clinic The largest complementary therapy clinic in the UK, The Hale Clinic has over 100 practitioners (nutritionists, naturopaths, healers, cranial osteopaths, acupuncturists etc), including Dr. Marilyn Grenville (see below) woman and baby specialist Dr Deborah McManners. All are based at 7 Park Crescent, London. Phone 020 7631 0156 (for bookings and brochures) 0870 850 2050 (for appts. with Dr. McManners) www.haleclinic.com.

Antony Haynes, The Nutrition Clinic Institute for Optimum Nutrition lecturer, Foresight practitioner and an increasingly well known nutritionist in the natural products industry, Antony founded The Nutrition Clinic at 89 Bryanston Court, London W1H to offer nutritional programmes for individuals to help them achieve optimum health. He specializes in all manner of ailments and sports nutrition. Phone 020 7723 3788.

Daphne Lambert, Penrhos School of Food and Health Qualified chef/nutritionist Daphne, based at Penrhos Court in Herefordshire, gives one-to-one consultations and runs residential and non-residential courses on vegetarian wholefood cookery (including a one day vegetarian Christmas course), organic food and health, women's health, sports nutrition and mother and baby care. She also co-organizes the Green Cuisine events held throughout spring, summer and autumn at Penrhos on organic food and health subjects and issues. She contributed to the Organic Food, Health & Nutrition Tips section in this book and is co-author of *The Organic Baby & Toddler Cookbook*. Phone 01544 230720 www.greencuisine.org daphne@greencuisine.org.

Maharishi Ayur-Veda Mother & Baby Programme A supportive programme through pregnancy, birth and afterwards, incorporating diet, massage for mother and baby, emotional and physical support. Based on the principle that the better the mother's health and general wellbeing, the better for the baby. Contact London Maharishi Ayur-Veda Health Centre in London Phone 020 7724 6267.

Dr Michel Odent, Primal Health Research Centre Dr Odent, the French natural childbirth pioneer, is a leading proponent of pre-conception care and the Accordion Method of losing contaminated fat well before conception. The Centre has a databank with 400 entries, now available to buy on disk, which explores the correlation between the primal period (birth to one year) and health later in life. The Centre is based at 59 Roderick Rd, London NW3 2NP. Phone 0207 485 0095 Modent@aol.com.

Offspring Home visits for new babies and toddlers. Marilyn Kaye, a nurse and qualified nutritionist with the ION, covers new baby and weaning problems, parenting skills, nutrition, resuscitation and safety in the home. Phone 0208 954 2222.

Preparing for Conception Residential workshops for couples which cover nutrition and herbs, feng shui, shiatsu, clearing emotional issues, avoiding environmental toxins and more. Teachers are Paul (feng shui practitioner) and Debra Norton (holistic birth teacher) with assisting practitioners. Phone 01305 266 156.

Michael van Straten Author, broadcaster, osteopath, acupuncturist and naturopath, Michael writes a weekly column in *Woman* magazine and for the *Daily Express*. He offers online health advice on www.thehealthyforum.com. He practices in Cheddington, near Leighton Buzzard, Bedfordshire. Phone 01296 668913.

OUTINGS & COURSES

Organic farms around the country are part of the Soil Association's Organic Farms Network These all are open to visitors so that people may see how real organic farms work. Many have areas and activities for children as well as selling their own produce. See the Soil Association website for a list of farms and links to their webpages or contact farmnet@soilassociation.org. See also the entry for Willing Workers on Organic Farms (page 187): this organisation has a long list of organic farms that take working visitors for weekends, and sometimes mid-week also. If you really want to get away from it all, look for a health spa (many of which serve organic food) at www.uk-healthfarmdirectory.co.uk. See *The Organic Directory* (from Green Books) for more details on organic restaurants, stores and attractions around the country.

Audley End Organic Kitchen Garden Part of the Henry Doubleday Research Association gardens, this redevelopment is in Saffron Walden, Essex. It includes a walled kitchen garden set within magnificent grounds, and a Vine House built in 1804 which contains grapes planted in 1917. As well as the gardens and Victorian buildings, there's a garden shop. Pre-booked groups are welcome. Phone 01799 522 842 www.hdra.org.uk.

Breadmatters Residential organic bread-making courses are run regularly throughout the year by the Village Bakery in Cumbria. Phone 01768 881515 for a programme.

Centre for Alternative Technology In Machynlleth in west Wales, this is Europe's largest eco-centre for alternative technology and energy sources, with an internationally renowned interactive display and education centre promoting practical ideas and information on sustainable technologies. The visitor complex in Powys, which attracts over 70,000 visitors each year, is on a seven acre site. A spectacular water-balanced cliff railway ferries visitors up to the site. There's an on-site shop and organic restaurant, accommodation in eco-cabins, many useful publications, and a full programme of residential courses and weekends throughout the year. Phone 01654 705950 (general info) 01654 705981 (education & courses) www.cat.org.uk.

Earth Balance Earth Balance is a visitor centre and working organic farm at West Sleekburn Farm, Bedlington, Northumberland. There are orchards, tidal salt marshes with curlew and the odd otter, beef cattle, milling wheat for the bakery and malting barley for the brewery. Try the technology trail demonstrating the potential of sustainable energy and providing an insight into biomass, wind, water and solar power. There's also a visitor centre with hands on exhibitions, a gift shop and an organic café. Open all year round, admission is £2 for adults, £1.50 for concessions or £8 for a family. Call ahead to find out what talks, tours and events are being held. Phone 01670 821 000.

Earth Centre Four hundred acres of South Yorkshire landscape have been turned into an environmental 'theme' park with fun learning for all the family: activities and events from April to November. There's also an environmentally friendly restaurant and shop where you can buy organic foods and goods. Entry is £4.50 for adults, £3.50 for children and concessions, with family tickets from £10. Visit by train: the Conisborough rail station is right beside the site, which is open seven days a week from 10am–6pm, April through November. The address is The Earth Centre, Denaby Main, Doncaster DN12 4EA. Phone 01709 513 933 www.earthcentre.co.uk.

Lower Shaw Farm This organic smallholding and B&B near Swindon, Wiltshire, offers fun and educational day or residential courses and activities all year round for the entire family (see Organic Events). There's also the opportunity to stay and eat for free during working weekends. Phone 01793 771 080.

Neal's Yard Courses Offers three foundation courses: aromatherapy, nutrition and natural medicine. Held during weekends and spread out over up to a year at the Chelsea Physic Garden in London with leading experts and some residential weekends. The Natural Medicine course is accredited by the University of Westminster. Phone 020 7627 1949 www.nealsyardremedies.com courses@nealsyardremedies.com.

Old Spitalfields Market What better way to spend a Sunday than at this large market, which has several organic food stalls selling produce, baked goods and prepared foods. There are also juice bars. Historic buildings date back to 1893. Find it opposite Liverpool Street Railway and Underground stations. There's also a healthfood shop at Spitalfields Sunday to Friday.

Penrhos Court Hotel/Penrhos School of Food & Health Based in a lovingly restored 700 year old medieval manor farm on the border of Herefordshire and Wales, Penrhos was the first registered organic restaurant in the UK. Accommodation is available all year round in en suite rooms with organic meals and organic wines. Penrhos chef/nutritionist Daphne Lambert runs one-to-one consultations and residential and non-residential courses on vegetarian wholefood cookery (including a one day vegetarian Christmas course), food and health, women's health, sports nutrition and mother and baby care. Penrhos hosts Green Cuisine events throughout spring, summer and autumn on organic food and health subjects and issues. Phone 01544 230 720 www.greencuisine.org.

Ryton Organic Gardens Based in Coventry, these gardens are the perfect venue for a family day out—as well as the different flower, vegetable and fruit gardens (including the Geoff Hamilton Paradise Garden) and gardening displays, there's a well stocked organic shop and organic restaurant. A small flock of rare breeds of sheep can be fed during summer months. There are also events for both serious and hobby gardeners, and for the entire family throughout the year (see Organic Events). Entry fee is £3 for adults and free for children and concessions, and guided tours are available. Phone 02476 303517 www.hdra.org.uk.

Sedlescombe Vineyard This organic vineyard in East Sussex is England's oldest, dating back 2000 years. It is open to the public (OFN member). There's a vineyard trail for tranquil walks, picnic areas, and a vineyard shop for tasting and buying organic wines, cider and fruit juices from the vineyard and around the world. Sedlescombe is open April to December throughout the week and weekends, and weekends only from January to March. For group visits, call ahead to book. Phone 0800 980 2884.

The Village Bakery The award-winning Village Bakery can be found in Melmerby, Penrith, Cumbria. There's a licensed restaurant (wines and beers only), a shop where you can buy all the bakery's organic breads, biscuits and cakes, and residential bread-making courses which you can attend throughout the year. The bakery is open 8.30am–5pm Monday to Saturday and 9.30am–5pm on Sundays, and they also run a mail order service. Phone 01768 881515.

Willing Workers on Organic Farms (WWOOF) This wonderful organisation arranges for people to help on organic farms (about 80 in all, throughout the UK). The idea is that you learn about organic farming and gardening, and help the farm at the same time. Board and lodging is provided. PO Box 2675, Lewes, Sussex BN17 1RB. Phone 01273 476 286 www.wwoof.org.

Yalding Organic Gardens The younger cousin of Ryton Gardens in Coventry, Yalding Organic Garden, Benover Rd, Yalding near Maidstone in Kent has flower, herb and vegetable gardens including a picturesque knot garden. Yalding is open from May to September, 10am–5pm from Wednesdays to Sundays and on Bank Holiday Mondays (except April and October, when it is only open during weekends). Entry is £3 for adults, free for children. Phone 01622 814650 www.hdra.org.uk enquiry@hdra.org.uk.

• Also consider city farms and community gardens—many of which are farmed organically and are open to the public. Contact the Federation of City Farms and Community Gardens (0117 923 1800 www.farmgarden.org.uk) to find one near you. There's also an annual city farms and community gardens festival.

EVENTS

The Fresh Network Lecture series & workshops Leading UK and international speakers regularly talk in venues around the country about the raw food diet and its benefits, how to prepare the most nutritious meals, and how to follow certain diets, and optimum health regimes, etc. Phone 0870 800 7070 www.fresh-network.com info@fresh-network.com.

Green Cuisine events Events such as talks by leading experts on organic food and health subjects and issues are held over weekends from spring to autumn at organic hotel/restaurant and mediaeval manor farm Penrhos Court in Herefordshire. Phone 01544 230720 www.greencuisine.org daphne@greencuisine.org.

Healing Arts Festival Held in London in November, this alternative therapy 'hands on' show is held at the Royal Horticultural Halls in London SW1. Phone 020 7371 9191.

The Living London Festival This is a large environmental festival held in Battersea Park each June or July, featuring some organic exhibitors, live entertainment, recycled clothing fashion shows and more. The event is run by Global Partnership (0207 924 0974).

Lower Shaw Farm Weekend fun and educational events at this unique organic smallholding near Swindon include willow sculpture and structure-making, learning about all types of native mushrooms and toadstools, celebrations for Easter, Halloween and Christmas, crafts and activities, circle dancing with masks, and recycling fabrics to make new things for the home. Day or residential courses and activities are held all year round for the entire family. Phone 01793 771 080.

The Mind, Body & Spirit Festival Held in London in May and June, this established consumer festival featuring over 100 exhibitors (including organic companies), talks, workshops, performances and demonstrations on spirituality, personal growth, healing and more. You can see organic products (including an organic fibres fashion show), and there's also an organic vegetarian restaurant. Phone 020 7371 9191.

Natural Products Europe & Organic Products Europe Two trade shows in London organised by Full Moon Communications.

Neal's Yard Remedies Workshop Series Several of the Neal's Yard many shops around the country run workshops on aromatherapy, homeopathy, herbs and other areas of natural medicine. Phone 0207 627 1949 to find out which store is closest to you, and when events are held. Phone 0207 627 1949 (customer services) www.nealyardremedies.com cservice@nealsyardremedies.com.

Olivers Wholefood Store Lecture Series Food and health talks and workshops are held in spring and autumn at Olivers Wholefood Store, Station Approach, Kew Village, West London. Olivers publishes its own in-store newsletter, listing coming events—ask to go on the mailing list or pick it up when you're there. Phone 0208 948 3990 everyone@oliverswholefoods.co.uk.

Open Organic Gardens Each year organic gardens throughout the country are opened to the public in June and August to promote organic gardening. They range from hobby fruit and vegetable gardeners' large backyards to extensive public gardens. For a programme of gardens near you and opening dates and times contact the Henry Doubleday Research Association on 02476 303 517. Admission prices and times vary depending on the venue. www.hdra.org.uk enquiry@hdra.org.uk.

Organic Food Awards Sponsored by the *Daily Mail*'s *You* Magazine and organized by the Soil Association, these annual prestigious awards held in autumn showcase and reward the best brands and products on the market. If you vote through *You* Magazine, you have the chance to win tickets to the awards. For information contact the Soil Association. Phone 0117 929 0661 www.soilassociation.org info@soilassociation.org.

Ryton Organic Gardens These gardens near Coventry are open to the public year round. Run by the Henry Doubleday Research Association, the fruit, vegetable and feature gardens have regular events for serious and hobby gardeners, and families. There is an extensive shop on-site with organic food, books and gardening equipment, as well as an organic restaurant which holds Saturday night candle-lit dinners. Entry fee is £3 for adults and free for children and concessions. Phone 02476 303 517 www.hdra.org.uk enquiry@hdra.org.uk.

OTHER USEFUL CONTACTS

Local groups of the Soil Association and HDRA Contact the Soil Association (0117 929 0661) and Henry Doubleday Research Association (02476 303517) to find an active local group near you. Interesting talks on organic food, farming, gardening and health issues, events, outings and newsletters are among benefits for joining.

The Osteopathic Information Service Find out the benefits of osteopathy for babies and young children. Phone 0207 357 6655 or check out the website at www.osteopathy.org.uk.

Royal Society for the Protection of Birds If you're a bird lover, join the RSPB and receive its magazine *Birds* and useful mail order catalogue for gifts and home needs. Phone 01767 680 551 www.rspb.org.

Starbound Enterprise Personal trainer and super-fit mum Michele Wilburn runs Starbound, which provides top quality Trimilin rebounders to improve lymphatic drainage, hasten detoxification, weight loss, improve general health and mental clarity. In her *Starbound* book Michele outlines the easy-to-follow lifestyle bounce workout and regime. Rebounders (small trampolines) can also be purchased from the Fresh Network (01353 662849) or Wholistic Research Company (01438 833100). New online shop including books and videos. Phone 0207 284 1918 www.starbounding.com Michele@starbounding.com.

Also available from Green Books

THE ORGANIC BABY & TODDLER COOKBOOK
Daphne Lambert & Tanyia Maxted-Frost

"Highly practical, with meal planners, recipes and tips"—*Home Healthcare*

A comprehensive but easy-to-follow nutrition guide for babies from weaning to toddlerhood (four to six months to three years old). It recommends a seasonal, mainly raw wholefood organic diet, emphasizing raw food in spring and summer and lightly cooked in autumn and winter, and advises on how to achieve optimum health for babies and toddlers. It includes the basic principles of good nutrition for mother and baby, information on why to eat organic, seasonal meal planners, recipes for meals and juices, tips on how to adapt the meals into the family routine, the ideal lunchbox, and more. **128pp 234 x 156mm ISBN 1 870098 86 2 £6.95 pb**

THE ORGANIC DIRECTORY
edited by Clive Litchfield

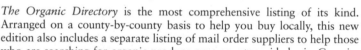

"With this excellent directory, life should now be easier for everyone, whether customers, growers, retailers or manufacturers. Unquestionably this is the most comprehensive guide yet to appear, a seriously sound job." —*Organic Gardening magazine*

The Organic Directory is the most comprehensive listing of its kind. Arranged on a county-by-county basis to help you buy locally, this new edition also includes a separate listing of mail order suppliers to help those who are searching for organic produce on a country-wide basis. Covering England, Scotland, Wales and the Channel Islands, in *The Organic Directory* you will find the names, addresses and phone numbers of: • Retailers, producers, wholesalers and manufacturers of organic food • Vegetable box schemes (weekly boxes of in-season vegetables from organic farmers) • Suppliers of organic gardening materials • Restaurants and accommodation specializing in organic food • and a wealth of other information including details of labelling schemes for organic produce; farm shops and farm gate sales; the WWOOF (willing workers on organic farms) movement; education opportunities; a sketch of the main organisations involved in the organic movement; a review of the potential dangers posed by genetic engineering; and a description of what terms such as organic and permaculture really mean. **2002/3 edition price £4.95**

BACKYARD COMPOSTING
John Roulac

Composting at home reduces your personal volume of rubbish, conserves water, increases plant growth, replaces the need for toxic chemical fertilizers and pesticides, and is also fun. With this little book you can learn how easy it is to: *START* Discusses all types of composting bins and how to build your own bin from scrap materials. *MAINTAIN* Includes easy-to-make 'hot recipes', time-saving tips and a troubleshooting chart. *USE* Save money by making your own free fertilizer at home from leaves, grass and kitchen scraps. **96pp with photos and line drawings ISBN 1 900322 11 0 £4.95**

Phone for our complete catalogue: 01803 863260
or visit our website: www.greenbooks.co.uk